Introduction to Biomolecular Energetics
Including Ligand–Receptor Interactions

Introduction to Biomolecular Energetics
Including Ligand – Receptor Interactions

Irving M. Klotz
Department of Chemistry
Northwestern University
Evanston, Illinois

1986

ACADEMIC PRESS, INC.
Harcourt Brace Jovanovich, Publishers
Orlando San Diego New York Austin
London Montreal Sydney Tokyo Toronto

COPYRIGHT © 1986 BY ACADEMIC PRESS, INC.
ALL RIGHTS RESERVED.
NO PART OF THIS PUBLICATION MAY BE REPRODUCED OR
TRANSMITTED IN ANY FORM OR BY ANY MEANS, ELECTRONIC
OR MECHANICAL, INCLUDING PHOTOCOPY, RECORDING, OR
ANY INFORMATION STORAGE AND RETRIEVAL SYSTEM, WITHOUT
PERMISSION IN WRITING FROM THE PUBLISHER.

ACADEMIC PRESS, INC.
Orlando, Florida 32887

United Kingdom Edition published by
ACADEMIC PRESS INC. (LONDON) LTD.
24–28 Oval Road, London NW1 7DX

Library of Congress Cataloging in Publication Data

Klotz, Irving M. (Irving Myron), Date
   Introduction to biomolecular energetics.

   Includes index.
   1. Bioenergetics. 2. Biomolecules. I. Title.
QH510.K57 1986   574.8'8   85-15750
ISBN 0–12–416262–2 (alk. paper)
ISBN 0–12–416263–0 (pbk.)

PRINTED IN THE UNITED STATES OF AMERICA

86 87 88 89   9 8 7 6 5 4 3 2 1

Dedicated to the memory of

Mark Kac

(August 3, 1914 – October 25, 1984)
whose lucid mind and generous spirit
served to enrich all who knew him

A theory is the more impressive the greater is the simplicity of its premises, the more different are the kinds of things it relates and the more extended is its range of applicability. Therefore, the deep impression which classical thermodynamics made upon me. It is the only physical theory of universal content which I am convinced, that within the framework of applicability of its basic concepts, will never be overthrown.

A. Einstein

Contents

Preface   *xiii*

Introduction: The Scope of Energetics   *1*

1. The Concept of Energy: The First Law of Thermodynamics   *3*

2. The Concept of Entropy   *9*

  A. The Second Law of Thermodynamics   *9*
  B. Entropy as an Index of Exhaustion   *13*

3. The Free Energy or Chemical Potential   *18*

  A. A Criterion of Feasibility of a Material Transformation   *18*
  B. Distinction between $\Delta G°$ and ΔG *21*

4. Computations of Standard Free Energies   *23*

  A. Calculations from Equilibrium Constants   *23*
  B. Calculations from Oxidation–Reduction Potentials   *28*

C. Calculations from Enthalpy and Entropy Changes   29
D. Calculations from Free Energies of Formation   32
Exercises   34

5. The Dependence of Chemical Potential on Concentration   37

A. Fundamental Relationship   37
B. Illustrative Calculations   39
Exercises   47

6. Group-Transfer Potential: High-Energy Bond   50

A. Comparison of Transfer Potentials   50
B. Coupled Reactions   56
C. Experimental Determination of Group-Transfer Potentials   59
Exercises   64

7. Some Laws of Physicochemical Behavior   68

A. Electrochemical Relationships   68
B. Osmotic Pressure   71
C. Molecular Weight from Ultracentrifugation   73

8. Energetics from a Molecular-Statistical Viewpoint   76

A. Fundamental Assumptions   77
B. Relationship to Thermodynamic Quantities   85
C. Some Applications   87

9. Energetics of Enzyme Kinetics   95

A. Concentration Profiles   96
B. Activation Energy Diagrams   100
Exercise   102

Contents

10. Ligand–Receptor Interactions *103*

    A. Multiple Equilibria *104*
    B. Graphical Representations *111*
    C. Analytical Representations of Data *118*
    D. Thermodynamic Quantities *128*
    Exercises *131*

Index *135*

Preface

With the increasing tendency to interpret biological phenomena in molecular terms, many biologists have become anxious to understand the principles of energetics which govern biochemical changes. Nevertheless, they have been understandably reluctant to devote the large amount of time that would be required even to peruse only a standard textbook in this field and to acquire the mathematical background upon which the logical development in such texts is based. However, for many purposes a "reading knowledge" of the language of thermodynamics may suffice, and this can be acquired without a large expenditure of time. With such knowledge one can at least understand the acknowledged classics in the field; some readers may even be stimulated to acquire the background to apply the concepts of energetics to branches of biology hitherto uninfluenced by these ideas.*

This paragraph, written in the 1950s, stated the essence of the philosophy of the first version of this small monograph, and it is still appropriate today. In the intervening three decades, the training of students in the sciences has become more intensive and quantitative so that even those at elementary levels are being introduced to concepts and some computations that were reserved for graduate work in chemistry a generation or two ago or for advanced studies in theoretical physics and mathematics a century ago. While retaining the original philosophy of this book, I have added several sections devoted to quantitative calculations and some topics that have become more prominent in the biochemical and related sciences in recent years. Thus I hope that a reader who has conscientiously worked his way

---

*From Klotz, I. M., "Some Principles of Energetics in Biochemical Reactions," Academic Press, New York, 1957, p.v.

through this volume will acquire not only a cocktail-party knowledge of thermodynamics but will be able to apply the principles to some biochemical or chemical transformations.

No doubt some astute reviewer will list a variety of topics not present in this small volume, or will point out that certain subjects ought to be treated more profoundly, and in fact are so treated in some other text, either of his or mine. The contents of a small monograph of this type obviously reflect the author's interests, taste, and judgment. Another teacher or investigator might select or emphasize different subjects. I can only hope that my selection will prove suitable for some students interested in acquiring an insight into the fundamental concepts and simple calculations of biochemical energetics.

Introduction: The Scope of Energetics

The objective of the field of energetics or thermodynamics is to establish the principles and laws which govern material transformations. Historically speaking, the subject was initially developed with a primary focus on energy transformations; special attention was directed toward heat engines. Hence the name thermodynamics was precisely descriptive of this branch of learning. Starting near the end of the nineteenth century, however, the dominant point of view changed to one that emphasized the development and use of energy functions to describe the state of a material system and to prescribe rules that govern transitions from one state to another. The energy functions are thus used as a method of bookkeeping in correlating the behavior of matter; hence the name energetics is perhaps more appropriate than thermodynamics to describe this field of knowledge.

There have been two general approaches to the field of energetics. The classical, or phenomenological, one is based on purely macroscopic observations and concepts; its fundamental ideas, although often highly abstract, can be reduced to common experience. The statistical-molecular approach, in contrast, assumes in addition that matter consists of discrete particles, molecules, whose behavior follows the laws of mechanics. By appealing to molecular images, the statistical viewpoint provides a more concrete mechanical interpretation for thermodynamic concepts. Being based on a molecular model, statistical thermodynamics can use experimental data to provide information on molecular properties. On the other hand, the classical theory in itself is unable to give any information on molecular characteristics. However, this limitation is in a sense also a virtue, for classical energetics is completely independent of any future changes in our conception of molecules. At least for macroscopic phenomena, classical energetics provides a completed logical framework. It is in regard to this corpus of physical theory that Einstein directed the observation printed after the title page of this volume.

We shall examine the subject of energetics first primarily from the phenomenological viewpoint since this approach requires a less sophisticated mathematical background. Once having grasped the basic principles and techniques, we shall also consider briefly some of the insights that are provided by statistical thermodynamics and molecular constructs, particularly when they are fused with classical energetics.

1

The Concept of Energy: The First Law of Thermodynamics

Historically, the concept of energy first arose in considerations of purely mechanical phenomena. It was recognized early in mechanics that if work were done to lift a weight from one position in the gravitational field to another (Fig. 1.1), this work could then be fully recovered. Thus for example, the weight resting on the table top in Fig. 1.1, if suitably connected to the weight on the floor at the right-hand side, could lift the latter to the top of the table.* Thus, one may assign to the first weight on the table top a certain "capacity to do work," which is latent in the weight by virtue of its position in the gravitational field. This capacity to do work, in the ideal situation of no friction, has the very special attribute of

---

* This statement is true strictly only for the limiting case in which there is no friction at any of the mechanical contacts, e.g., between the wheel and its bearing in Fig. 1.1. Likewise, it is to be understood that the weight on the left in the figure is a fraction of a milligram heavier than that on the right.

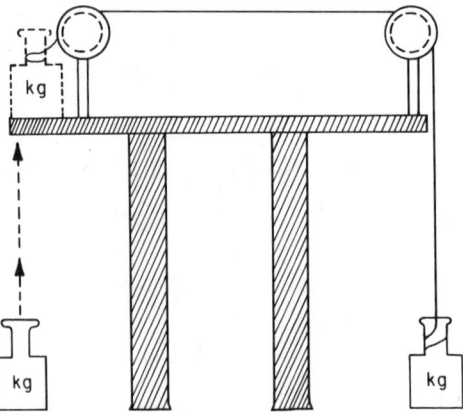

Fig. 1.1. Energy and work in a purely mechanical system: a gravitational field.

depending only on the height of the weight above the floor and not on the path taken to bring the weight from the floor to the table top. It seems appropriate, therefore, to assign a special name, *energy*, to this capacity to do work, and perhaps to add the adjective *potential* to the word *energy* to emphasize the latent character of the work capacity in this particular example.

Work expended in causing extension of a spring is also an early example of energy in the mechanical sense (Fig. 1.2). If x represents the extension of a spring at rest, then a certain quantity of work must be exerted on the spring to increase it extension to x'. Again this work may be recovered (fully in ideal circumstances), for example, by attaching the extended spring to a rope and pulley system like that shown in Fig. 1.1 and raising the weight on the right-hand side. Again the capacity of the ideal spring to do work depends only on the distance of extension x', and hence we again may assign the special name *energy* to this capacity to do work.

One might be tempted to generalize from these two examples and think that in all cases in which one exerts a force and does some work on a body, the body in its new position retains its capacity to do work. A very familiar example would immediately convince us that this generalization cannot be maintained. In Fig. 1.1, for example, a substantial amount of work would be required to move the weight along the floor (particularly if the floor

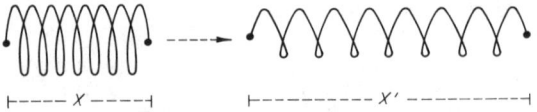

Fig. 1.2. Energy and work in a purely mechanical system: a spring.

were very rough) from the left side to the right side of the table. In this case, the weight on the right does not retain the capacity to do work. In fact, one must do work on it to return the weight to its initial position. Clearly in this situation the exertion of work on a body does not result in "giving it energy."

Since no modern student approaches this subject without some school exposure to physics, he or she will quickly point to an important distinction between our third experiment with the weight and the first two experiments. In the former, in which friction is dominant, heat is evolved as the weight is moved along the floor; without hesitation the student will inform us that heat is a form of energy. It may come as a shock to him or her to be told that the thought of fusing heat and work into the unified concept of energy occurred solely to a few scientific geniuses and only a little over a hundred years ago, in the first half of the 19th century.

Before that period, and particularly in the later part of the 18th century, when quantitative calorimetry was invented, heat was viewed as a totally separate, independent natural entity—caloric. Caloric is an eminently attractive commonsense concept for rationalizing simple observations such as temperature changes when a cold body is placed in contact with a hot one. The cold body appears to have extracted something from the hot one. Furthermore, if both bodies are constituted of the same material and the cold object has twice the mass of the hot one, then one observes that the increase in temperature of the former is only half the decrease in number of degrees of the latter. Thus, a conservation principle, conservation of caloric, arises naturally. From that, the notion of flow of a substance from the hot to the cold body appears almost intuitively, together with the concept that the total quantity of the caloric removed from or added to a body can be represented by the product of the mass multiplied by the temperature change. With these ideas, Black was led to the discovery of specific heat, heat of fusion, and heat of vaporization. Such successes established the concept of caloric so solidly and persuasively that it blinded the eyes of even the greatest scientists of the 18th and early 19th centuries. Thus, they did not see well-known facts that were common knowledge even in primitive cultures—heat can be produced by work against friction. Ultimately even theoretical physicists recognized that heat and work are intertwined.

We should realize, however, that in designating heat as a form of energy, we have definitely departed from our initial definition that energy is the capacity to do work. Furthermore, there would be no advantage in making heat comparable to work unless the two were quantitatively interrelated. As every student knows, however, this quantitative equivalence was demonstrated in the experiments of Joule.

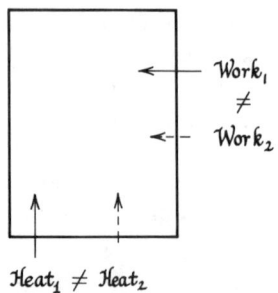

Fig. 1.3. Relationships between work and heat according to the first law of thermodynamics.

Let us now leave behind this brief historical excursion and adopt a more sophisticated general analysis of the relationships between work, heat, and energy. For this purpose, we may represent schematically (Fig. 1.3) any one of the many physical or chemical systems with which we shall be concerned. In the special circumstance illustrated, it is assumed that the system has work done on it from the outside and absorbs heat from the surroundings; alternative circumstances would be represented with arrows pointing in the reverse direction.

As one example of the system represented by Fig. 1.3, let us consider a heavy boulder resting on the edge of a steep cliff. This boulder could be brought to the floor of the valley in many ways. One method would be to attach it to one end of a rope, attach another boulder of almost equal weight, resting at the bottom of the valley, to the other end of the rope and by stringing the rope over a pulley wheel suitably mounted at the brink of the cliff, allow the high boulder to be lowered to the ground while the second boulder is raised to the top of the cliff. In this procedure the first boulder performs work on something outside of it but no heat is absorbed or evolved in the process. Another method for getting the boulder down might be to let it slide down some pathway to the valley. Less work could be obtained by this method, but a definite quantity of heat would be evolved. Many other paths could be chosen differing in the frictional resistance they would offer to the motion of the boulder. For each path, a different amount of work is performed, and a different quantity of heat is evolved.

To maintain a self-consistent bookkeeping record, we agree that if work W is done *on* the system, it will be assigned to positive numerical value, and if heat Q is absorbed *by* the system, it will be assigned a positive numerical value. Examining the individual values of W and Q for the first boulder in each of the different paths traversed, we find that despite great variations in individual values of W and large differences in values of Q, the *sum $Q + W$ in all experiences* is the same, so long as the respective starting point and final resting point of the boulder's travel are the same for each path.

As another example of Fig. 1.3, we might consider a stretched rubber band. The band could be attached to a weight, so that as the band contracts it does some work W_1. In this process, a certain amount of heat Q_1 would be absorbed. A second "path" for the return of the stretched rubber band to a shorter length might be to just let it snap back, i.e., to let $W_2 = 0$; in this process, there would be an associated heat effect Q_2 that is different from Q_1. Once again, though, despite the inequality of W_1 and W_2 or Q_1 and Q_2, the *sum $Q_1 + W_1$* has been found to be equal to $Q_2 + W_2$, for a wide variety of different paths between the same end points.

This special character of the sum $Q + W$ is not limited to mechanical processes. Consider, for example, the chemical reaction

$$\text{Pyruvate} + H_2 \rightarrow \text{Lactate.} \tag{1.1}$$

This reaction can take place in a suitable electrochemical cell so that electrical work is obtained as the reaction proceeds. Under certain ideal conditions the work performed by the electrochemical cell on some outside system is 11,440 cal; therefore, W (work done on the electrochemical cell) must be assigned the value $-11,440$ cal. In this ideal process, 10,200 cal of heat are evolved by the cell. Therefore Q is $-10,200$ cal and $Q + W = -21,640$ cal. Under certain nonideal conditions the electrochemical cell may perform 10,000 cal of work ($W = -10,000$ cal); in this process, 11,640 cal of heat are evolved by the chemicals in the cell, so Q is $-11,640$ and $Q + W = -21,640$ cal. In contrast, the same chemical reaction can be carried out in a closed vessel without any work being performed ($W = 0$); under these circumstances, 21,640 cal of heat are observed to be evolved ($Q = -21,640$; $Q + W = -21,640$ cal). Thus, we see that $Q + W$ is the same for all three methods (or paths) for carrying out the chemical reaction shown in Eq. (1.1).

We have considered three special cases of the schematic diagram of Fig. 1.3. Going much further, we can summarize a vast amount of experience by generalizing the statements we have made for Q, W, and $Q + W$ so far only in our three special examples. In any system we find that although the value of W, the work done on the system, or Q, the heat absorbed by the system in going from one state to another varies with the path chosen, the *sum $Q + W$* is invariant and depends only on where we start and where we end. This remarkable characteristic of the quantity $Q + W$ may be emphasized by giving it a special name—the change in energy* ΔE of the system. In terms of a simple equation we may write

$$\Delta E = Q + W. \tag{1.2}$$

The essence of the first law represented by Eq. (1.2) thus may be summarized in two statements. The first defines a new concept, energy, in terms of directly measurable concepts, heat and work. The second declares that the internal energy, so defined, is a thermodynamic property; that is, it depends only on the state of a system and not on the previous history of the system. To repeat, despite the fact that the heat absorbed Q and the work absorbed W in going from one state to another depend on the particular path used in the transformation, the sum of the two quantities, defined as ΔE, is independent of the method by which the change is accomplished.

Furthermore, it is a necessary consequence of these statements that ΔE around a closed path is zero. If the system is ultimately returned to its initial state, there can be no change in its energy, since E depends only on the state of the system, which has not changed in a cyclic process. Despite this fact, the total Q or total W for the cyclical process may differ very much from zero.

It should also be noticed that the very definition of the energy concept precludes the possibility of determining absolute values; that is, we have defined only a method of measuring changes in internal energy in terms of heat and work quantities. Classical energetics, by itself, is incapable of determining the absolute zero of reference for the energy function. In practice, this limitation is not really a handicap, however, since interest is generally focused on chemical and physical transformations, and any convenient state may be chosen as the reference point.

---

* We shall use the symbol E for the internal energy of a system. The first law of thermodynamics permits us to define, however, only changes in energy $E_2 - E_1$, where E_2 is the energy of the system at the end of a process and E_1 that at the beginning. For $E_2 - E_1$ we shall use the briefer notation ΔE.

2

The Concept of Entropy

A. THE SECOND LAW OF THERMODYNAMICS

We are interested in the principles of energetics primarily for their ability to tell us whether a particular biochemical (or biophysical) change is feasible under a specified set of conditions. The first law of thermodynamics by itself, however, does not provide us with a criterion for determining when a transformation may occur spontaneously, as a few specific examples will show.

One unfamiliar with energetics tends to believe that a chemical or physical change can occur spontaneously only if the final state of the system is at a lower level of energy than the initial (i.e., if ΔE is negative). This conclusion is based on thoroughly valid common experience with movements in gravitational fields. We all recognize that a ball tends to fall from a higher to a lower level, not vice versa. For purely mechanical processes, involving no heat exchange, the rule that ΔE must be negative is valid as a criterion of permissible spontaneous change. What is not often realized is that this rule is completely unreliable if applied to all kinds of transformations.

There are spontaneous transformations that occur despite the fact that the internal energy of the system at the end is essentially the same as before the transformation. For example, when a stretched rubber band is released, it snaps back spontaneously: yet ΔE for this process is substantially zero.

There are even processes that can occur spontaneously despite the fact that the internal energy at the end is greater than at the beginning of the transformation (i.e., for which ΔE is positive). For example, if an aqueous solution of methane is placed in contact with benzene, the methane will move spontaneously from the water to the organic solvent; yet ΔE for the transfer is +2.8 kcal/mole. Positive values of ΔE are particularly common among chemical changes. For example, when a free copper ion and a protein molecule (P) are permitted to come in contact with each other, they will spontaneously form a complex compound:

$$P + Cu^{2+} \rightarrow P\text{--}Cu^{2+}. \tag{2.1}$$

Nevertheless, for this reaction ΔE is +3 kcal; that is, P–Cu^{2+} has 3 kcal more internal energy than separated P and Cu^{2+} particles. An even more familiar example is the ice–water phase transition. At any temperature near 0°C liquid water is in a higher energy state than ice; nevertheless if the ice were at 1°C it would spontaneously be transformed to water.

Thus, reactions may occur spontaneously for which ΔE is negative, zero, or positive (Fig. 2.1). Clearly ΔE is no criterion of permissibility of a transformation. Apparently the first law of thermodynamics does not contain within it the basis of any criterion of spontaneity. Some further principle will be required to summarize in a general statement the observed tendency of systems of many different types to change in a particular direction.

Let us examine again a process of the type shown in Fig. 2.1c, that is, one that proceeds spontaneously to a state of higher energy. At first glance, the occurrence of this transformation might seem to violate the first law, for how can a system spontaneously acquire energy? Actually, of course, this system does not violate the first law; during the spontaneous transformation it absorbs heat (3 kcal in Fig. 2.1c, 59 kcal in Fig. 2.1h) spontaneously from its surroundings.

Evidently then, some chemical systems are capable of spontaneously absorbing heat from their surroundings in order to proceed to a higher energy state whereas others are not. Is there any way of telling which systems will behave in which way?

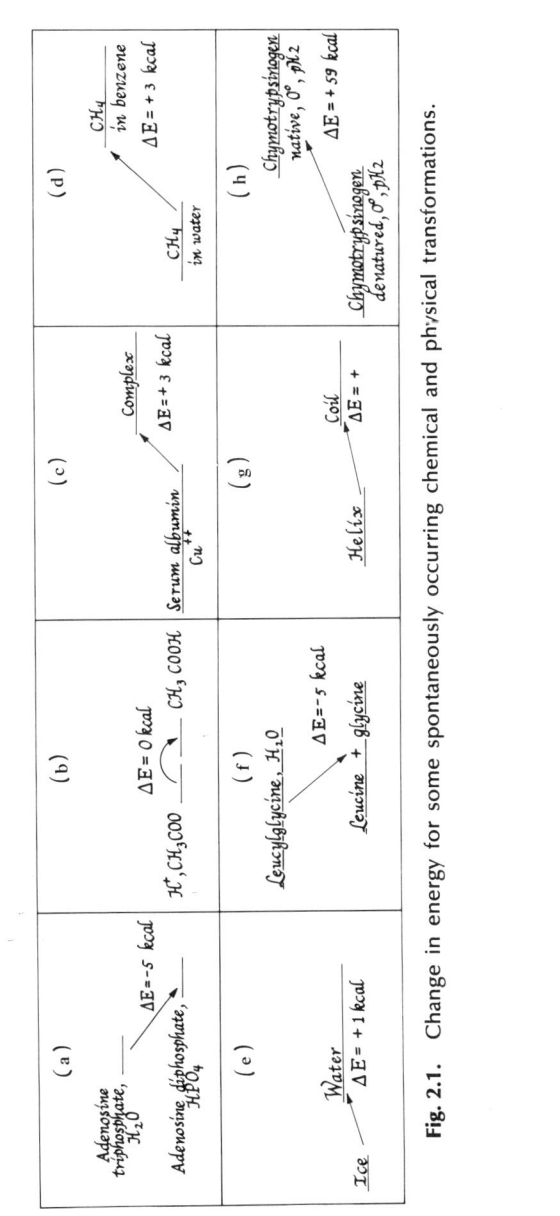

Fig. 2.1. Change in energy for some spontaneously occurring chemical and physical transformations.

TABLE 2.1

Intensive and Extensive Factors in Forms of Energy

	Intensity	Capacity
Work; gravitational	Height h	Mass m
Work; stretching	Tension τ	Length l
Work; electrical	Voltage ε	Charge q
Work; expansion	Pressure P	Volume V
Work; surface	Surface tension γ	Area A
Heat	Temperature T	?

Since ΔE as a criterion of spontaneity fails for transformations involving heat exchange, we might try another thermodynamic quantity, for example, Q. Clearly the proper criterion does not depend directly on Q, since reactions proceed spontaneously for which Q is positive, negative, or zero. It has turned out, nevertheless, largely as a result of the insight of two great physicists of the 19th century, Kelvin and Clausius, that a suitable rule can be developed based on a modified function of Q instead of Q itself.

We can make a reasonable guess as to the form of this function by considering certain analogies between energy in the form of work and energy in the form of heat.

When we analyze in more detail a variety of kinds of work, we find in each case that the total work is determined by two factors, an intensity factor and a capacity factor (Table 2.1). In doing work against gravity, the total work depends on the change in height h and also on the mass m of the material moved. Electrical work is determined by the voltage drop as well as by the total charge moved by (or against) a particular voltage drop. When a gas expands and does work against a confining piston, the work done depends on the pressure of the gas and also on the volume change during the working expansion.

Insofar as heat energy is concerned (Table 2.1), it is clear that the intensity factor should be the temperature. What, however, should be the corresponding capacity factor?

By a simple dimensional analysis we can see that the quantity Q/T would, at least, have the proper form:

Energy = Intensity Factor × Capacity Factor,

$$Q(\text{cal}) = T(\text{deg}) \times \frac{Q(\text{cal})}{T(\text{deg})} \tag{2.2}$$

It would clearly be desirable to see whether this new quantity shows any correspondence with spontaneity in material transformations.

B. Entropy as an Index of Exhaustion

We can summarize a long history of intellectual effort by saying that the new function, if measured in certain prescribed ways, does indeed provide the criterion we have been seeking. The new criterion found may be described as follows. If we wish to know whether a particular system can change spontaneously from one state a to another b, then, in principle, we must evaluate Q/T for a transformation over the same states but carried out in a reversible* manner. Furthermore, we must also evaluate Q'/T for any changes that would occur in the surroundings during the actual spontaneous transformation. As a result of much experience we know that under natural conditions heat flows spontaneously from a hotter object to a colder one. If we accept this observation as an empirical law of nature, then a remarkable equation can be proved to be *true for any transformation that can occur spontaneously*,

$$(Q/T)_{\text{system}} + (Q'/T)_{\text{surroundings}} = \text{a positive number.} \qquad (2.3)$$

For simplicity in notation we may write also

$$\sum (Q/T) > 0 \qquad (2.4)$$

as the criterion of spontaneity. If

$$\sum (Q/T) = 0, \qquad (2.5)$$

then the system under consideration is at equilibrium.[†]

As in the case of the quantity $(Q + W)$ discussed in Chapter 1, so here the function (Q/T) evidently has very unique characteristics in being able to predict the behavior of physical and chemical systems. It seems appropriate, therefore, to endow it with a special name, the change in *entropy*, and attach to it a special symbol ΔS. We may replace Eq. (2.4) and (2.5), therefore, by the following statements, to conform with the forms more commonly found in the scientific literature:

$$\sum \Delta S > 0, \quad \text{spontaneous change possible;} \qquad (2.6)$$

$$\sum \Delta S = 0, \quad \text{system at equilibrium.} \qquad (2.7)$$

B. ENTROPY AS AN INDEX OF EXHAUSTION

To obtain a better grasp of the essential character of the entropy concept let us examine in more detail a few transformations that occur spontaneously despite the fact that $\Delta E = 0$ in each case (Fig. 2.2).

---

* For a discussion of the strict definition of "reversible," see any standard text on thermodynamics.

† If $\Sigma(Q/T) < 0$, then for the reverse transformation $\Sigma(Q/T) > 0$. Hence, the *reverse* transformation can be spontaneous.

If two blocks of metal, one at a high temperature T_3 and the other at a low temperature T_1 (Fig. 2.2) are separated by a perfect heat insulator and the system as a whole is surrounded by a thermal blanket that permits no transfer of heat in or out, then there can be no change in internal energy with time. If the insulator between the blocks is removed, the hotter block will drop in temperature and the cooler one will rise in temperature until the uniform temperature T_2 is reached. This is a spontaneous transformation; ΔE is 0. However, there has been a loss of *capacity to do work*. In the initial state, one could insert a thermocouple in the block at T_3 and another in that at T_1 and obtain electrical work. At the conclusion of the spontaneous transformation (without the thermocouples) the internal energy is still the same as that at the outset, but it is no longer in a condition where it has the capacity to do work.*

Similarly if two ideal gases at a high pressure P_3 and a low pressure P_1, respectively, are separated by a partition, the high pressure gas will move spontaneously into the low pressure chamber if the separating barrier is removed (Fig. 2.2b); yet, $\Delta E = 0$. Again there has been a loss of *capacity to do work*. In the initial state, one could arrange a piston with a rod extending to the outside of the containers, have the gas at P_3 on one side of the piston and the gas at P_1 on the other and the piston could perform work (e.g., by lifting a weight against gravity). That capacity to do work is no longer present at the conclusion of the spontaneous transformation (with no piston present); the character of the energy has changed.

A similar analysis can be made for two solutions with a solute (e.g., an electrolyte such as $CuSO_4$) at two different concentrations being allowed to mix spontaneously upon removal of a separating partition (Fig. 2.2c). A copper metal electrode inserted into solutions C_2'' and C_2' will generate electrical energy, but such a pair of electrodes in the same physical position after spontaneous mixing has been completed to give solution C_2 can produce no work. In a very different type of system, a stretched rubber band (Fig. 2.2d), release of the constraint will spontaneously lead to the

---

* The energy is "differentiated" in the separated bodies and "dedifferentiated" after thermal equilibration. The entropy is different in the differentiated state of this system than it is in the dedifferentiated state; the entropy is an index of the extent of dedifferentiation. Corresponding statements can be phrased for two gases initially at different pressures or two solutions with different concentrations of Cu^{2+} (Fig. 2.2). In a molecular visualization, a system also has a larger entropy when it is dedifferentiated. For example, let us place three layers of black balls at the bottom of a cubic box and carefully place three layers of white balls on top of the black layers. If then the box is allowed to be buffeted around, in time the balls will achieve one of many possible random arrangements of black ones and white ones. At the outset, the system was very differentiated; at the end of the transformation, it reached a highly dedifferentiated state, one which Boltzmann associated with a larger entropy (see Chapter 8).

B. Entropy as an Index of Exhaustion

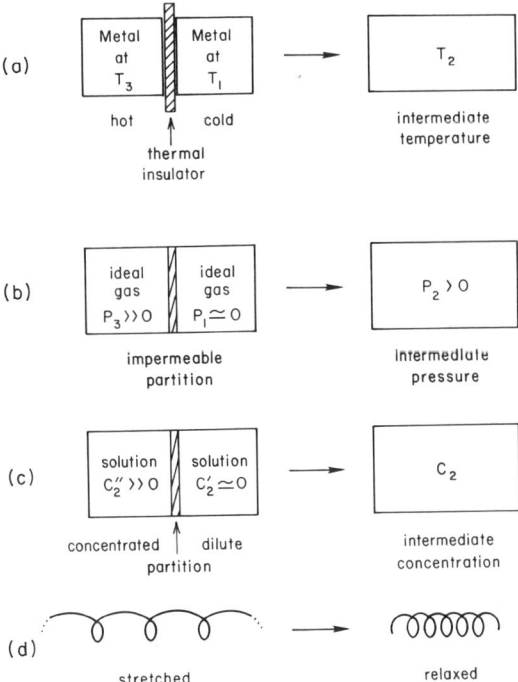

Fig. 2.2. Some spontaneous, irreversible transformations for which $\Delta E = 0$. In each case there has been a loss in capacity to do work.

relaxed rubber band. In each system $\Delta E = 0$; but again the capacity to do work has been diminished.

This loss in capacity to do work is a property of each of the systems illustrated in Fig. 2.2, whether or not it has actually done any work during the transformation.

Thus, we ought to view entropy as an index of condition or character of a system (perhaps somewhat analogous to a cost-of-living index or to pH as an index of acidity). It is an index of the state of differentiation of the energy, an index of the capacity to do work, an index of the tendency toward spontaneous change. The more a system exhausts its capacity for spontaneous change, the larger the entropy index. Hence, we should preferably say that *entropy is an index of exhaustion*: the more a system has lost its capacity for spontaneous change, the more this capacity has been exhausted, the greater is the entropy.

Thus, the second law of thermodynamics provides us with a new function, the entropy change ΔS to be used as the fundamental criterion of spontaneity. For a closed region of space (for which, therefore, $\Delta E = 0$) including all changes under observation,

$$\Delta S \geqq 0, \tag{2.8}$$

the equality sign applying to systems at equilibrium, the inequality to all systems capable of undergoing spontaneous changes.

Spontaneous transformations occur all around us all the time. Hence ΔS, for a section of space encompassing each such transformation and its affected surroundings, is a positive number. This realization led Clausius to his famous aphorism:

> Die Energie der Welt ist konstant:
> die Entropie der Welt strebt einem Maximum zu.*

To a beginning student this form of statement is frequently the source of more perplexity than enlightenment. The constancy of energy causes no difficulty, of course. Since energy is conserved, it fits into the category of concepts to which we attribute permanence. In thought, one usually picture energy as a kind of material fluid, and hence even if it flows from one place to another, its conservation may be visualized readily. However, when we carry over an analogous mental picture to the concept of entropy we immediately are faced with the bewildering realization that entropy is "being created out of nothing" whenever there is an increase in entropy in an isolated system undergoing a spontaneous transformation.

The heart of the difficulty of "understanding" the concept of increase in entropy is a verbal one. It is very difficult to dissociate the unconscious verbal implications of a word that we have used all of our lives in other contexts without critical analysis. In speaking of "increase in entropy" we are using language appropriate for the description of material bodies. Automatically, therefore, we associate with entropy other characteristics of material bodies that are at variance with the nature of entropy and, hence, are a source of confusion.

Ultimately, one must realize that entropy is essentially a mathematical function. It is a concise function of the variables of experience, e.g., temperature, pressure, and composition. Natural processes tend to occur only in certain directions, that is, the pressure, temperature, and composition variables change only in certain ways, but very complicated ways, which are most concisely described by the change in a single function, the entropy function ($\Delta S > 0$).

Some of the historical reluctance to assimilate the entropy concept into general scientific thinking (and much of the introductory student's bewilderment) might have been avoided if Clausius had defined entropy

---

* The energy of the universe is constant; the entropy tends toward a maximum.

B. Entropy as an Index of Exhaustion

(as would have been perfectly legitimate to do) as

$$\sum \Delta S' = \sum (-Q/T) \tag{2.9}$$

with a negative sign instead of the positive one of Eq. (2.4). With this definition all of the thermodynamic consequences that have been derived from the entropy function would be just as valid except that some relations would change in sign. Thus, in place of Eq. (2.8) we would find that for an isolated system,

$$\Delta S' \leqq 0, \tag{2.10}$$

the equality sign applying to systems at equilibrium, the inequality to systems capable of spontaneous changes. Now, however, we would recognize that for all isolated sections of space undergoing actual changes, ΔS is a negative number; that is, the entropy decreases. Likewise, paraphrasing Clausius, we would say, "Die Entropie der Welt strebt einem *Minimum* zu." This statement would accord more obviously with our experience that observable spontaneous changes go in the direction that *decreases* the capacity for further spontaneous change and *decreases* the capacity to do work, and that the universe (or at least the solar system) changes in time toward a state in which (ultimately) no further spontaneous change will be possible. We need merely reiterate a few examples (Fig. 2.2): solutes always diffuse from a more concentrated solution to a dilute one; clocks tend to run down; magnets become self-demagnetized; heat always flows from a warm body to a colder one: gases always effuse into a vacuum; aqueous solutions of NaCl and $AgNO_3$ if mixed always form AgCl. Although some of these individual changes can be reversed by some outside agency, this outside agent must itself undergo a transformation that decreases its capacity for further spontaneous change. It is impossible to restore every system back to its original condition. On earth, our ultimate sources of energy for work are the sun or nuclear power; in either case, these ultimate nuclear reactions proceed unidirectionally and toward the loss of capacity for further spontaneous change.

In some respects, especially pedagogical ones, it might have been better to change the sign of the original definition of the index so that it would measure residual capacity rather than loss of capacity. However, with the development of molecular-statistical energetics (see Chapter 8) and the identification of entropy (in terms of kinetic–molecular theory) with the degree of disorder of a system, the original sign chosen by Clausius turns out to be the more felicitous one. The universal tendency of all changes to reduce everything to a state of equilibrium may be correlated with the rearrangements of molecules from orderly to disorderly configurations; since there are more disorderly arrangements than orderly ones, it is appropriate that the entropy index increase with the approach of all things to a state of equilibrium.

3

The Free Energy or Chemical Potential

A. A CRITERION OF FEASIBILITY OF A MATERIAL TRANSFORMATION

In principle, the second law of thermodynamics summarizes our experience, in terms of the properties of the entropy function, as regards the conditions under which a transformation can take place spontaneously. In practice, however, this function is not a convenient one to use. It requires us to evaluate the changes in entropy both in the changing substance and in its surroundings, in other words, in a section of space large enough to include all changes caused by the reaction. It would be much more convenient if we could restrict our attention merely to the material transformation itself, whose environment we can generally control, without paying any attention to changes in properties of the surroundings. Furthermore, it would be desirable to have a criterion of feasibility applicable under conditions normally characteristic of biological reactions: constant pressure (usually that of the atmosphere) and constant temperature.

A. A Criterion of Feasibility of a Material Transformation

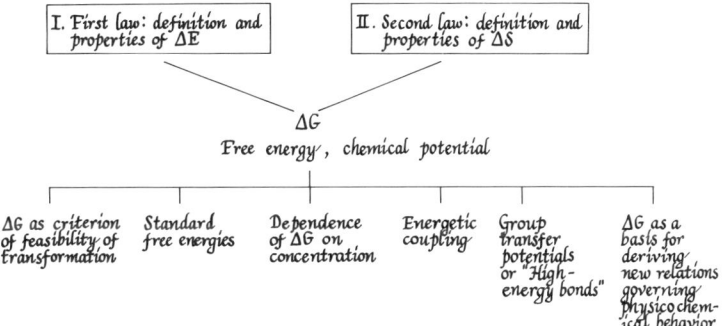

Fig. 3.1. Logical relationships in energetics.

These objectives were accomplished by Gibbs and by Helmholtz who created the *free energy function* by fusing the first and second laws into a single mathematical statement (Fig. 3.1). It was recognized that the law of conservation of energy requires that

$$(Q'/T)_{\text{surroundings}} \equiv \Delta S_{\text{surroundings}} = -\Delta(E + PV)_{\text{system}}/T, \quad (3.1)$$

where P and V are the pressure on and the volume of the system, respectively. Thus, the ΔS of the surroundings can be replaced by a function of E, P, V and T *of the system* undergoing the transformation. Substitution of the expression in Eq. (3.1) into Eqs. (2.3)–(2.7), followed by simple algebraic manipulation, leads to the following rule for a system at constant pressure and temperature:

$$\Delta E_{\text{system}} + P \Delta V_{\text{system}} - T \Delta S_{\text{system}} \leqq 0. \quad (3.2)$$

In this expression, every term is measured for the system itself. The inequality sign applies to any spontaneous change and the zero to any system at equilibrium. Now that each thermodynamic quantity in the criterion of feasibility refers to the system itself and the properties of the surroundings have been excised, we can omit the subscripts in Eq. (3.2). Furthermore, for conciseness we create a new function, the *Gibbs free energy* G,* defined so that Eq. (3.2) may be replaced by

$$\Delta E + P \Delta V - T \Delta S = \Delta G \leqq 0. \quad (3.3)$$

Thus, the free energy change for a specified reaction may be negative (in which case the reaction can proceed spontaneously) or zero (in which case

---

* The letter F has also been commonly used for the Gibbs energy, particularly in the United States. Many tabulations of chemical thermodynamic data, particularly older ones, use F (and, hence, ΔF).

the substances are at equilibrium).* Of course, ΔG could turn out to be a positive number, but in this case the exactly opposite reaction is accompanied by a negative ΔG and, therefore, this opposite reaction can proceed spontaneously.

In classical energetics, the first and second terms on the left-hand side of Eq. (3.3) are combined into another quantity, ΔH, called the change in enthalpy. Thus, an equivalent form of Eq. (3.3) is

$$\Delta G = \Delta H - T\Delta S. \tag{3.4}$$

For our purposes, however, it will usually not be necessary to consider the distinction between ΔH and ΔE. For most reactions that occur in solution, especially those of biological interest, the $P\Delta V$ term makes a negligible contribution to the energetics because the volume change ΔV during the course of the transformation is very small. Hence, for most practical purposes ΔH and ΔE will be indistinguishable in situations of primary interest to us, and thus we may write

$$\Delta G \cong \Delta E - T\Delta S. \tag{3.5}$$

As is evident from Eq. (3.5) the sign and size of ΔG will reflect contributions from both the change in internal energy and the change in entropy during the transformation. It is easy enough to accept the conclusion that if the internal energy of a system drops, i.e., if ΔE is negative, then ΔG should be negative, and the free energy drop should be available for useful work. However, it is not so simple to see at this point (since there exists no mechanical analogy) that even if ΔE were zero, ΔG could still be negative and useful work could be obtained; such is the case when ΔS is a positive number for a particular transformation. We shall consider in a later chapter some molecular properties related to ΔS and thereby obtain a greater insight into the significance of entropy changes. For the moment, let us merely examine one simple specific example of the counterplay of the ΔE and ΔS factors.

Liquid water may be converted to ice at different temperatures:

$$H_2O(l) \rightleftharpoons H_2O(s). \tag{3.6}$$

The changes in thermodynamic properties for this transformation are known in the temperature range slightly below and slightly above 0°C. These are summarized in Table 3.1. We note first that at each temperature there is a *drop* in energy as we go from the liquid to the solid; on this basis

---

* At constant pressure and temperature, G (or ΔG) depends only on the composition of the system and not on its history. This feature, plus its fundamental character of being a criterion of feasibility, makes the name "chemical *potential*" particularly appropriate.

B. Distinction Between $\Delta G°$ and ΔG

TABLE 3.1
Thermodynamic Properties of Water
$H_2O(l) \rightleftharpoons H_2O(s)$

Temperature (°C)	ΔE (cal/mole)	ΔH (cal/mole)	ΔS (cal/mole deg)	$-T\Delta S$ (cal/mole)	ΔG (cal/mole)
−10	−1343	−1343	−4.9	1292	−51
0	−1436	−1436	−5.2	1436	0
+10	−1529	−1529	−5.6	1583	+54

alone, water should spontaneously freeze at all temperatures between −10 and +10°C. In fact, the drop in energy is greatest at +10°C, and we might even expect the liquid to freeze most readily at this temperature, whereas actually it cannot do so at all. The counterbalancing factor is the entropy, which is negative at each temperature, and increasingly so as the temperature rises. (As we shall see in Chapter 8), a decrease in entropy is associated with a transformation to a more orderly state, and of course in this example it is obvious that crystalline ice contains a more orderly arrangement of molecules than does liquid water.) At −10°C, ΔS (when multiplied by T) contributes +1292 cal; this being numerically less than the ΔH (or ΔE) of −1343 cal, we arrive at a net ΔG of −51 cal/mole, and hence conclude that the transformation can occur spontaneously. At 0°C, ΔH (or ΔE) and $T\Delta S$ balance exactly and ΔG is zero; thus, the system is at equilibrium. As the temperature increases further, however, $-T\Delta S$ increases more rapidly than ΔH (or ΔE) drops, the net ΔG becomes a positive number, and we conclude that the *reverse* reaction (ice → water) must be the spontaneous one.

B. DISTINCTION BETWEEN $\Delta G°$ AND ΔG

We now have in the free energy function ΔG a convenient fusion of the first and second laws which provides a quantitative index of the potential ability of a substance to undergo a chemical or physical transformation. Before embarking on some actual numerical computations, however, it is advisable to clarify one additional feature of ΔG. Merely specifying the nature of the chemical change does not lead to a specific value for ΔG; the magnitude also depends on the conditions under which the reaction is carried out.

For example, let us consider the synthesis of the following peptide bond from aqueous solutions of alanine (Ala) and glycine (Gly):

$$\text{Ala (aq)} + \text{Gly (aq)} \rightleftharpoons \text{Ala—Gly (aq)} + H_2O(l). \quad (3.7)$$

If the reaction occurs at the concentrations:

$$\text{Ala(aq)} + \text{Gly(aq)} \rightleftharpoons \text{Ala—Gly(aq)} + \text{H}_2\text{O(l)},$$
$$1\,M \qquad 1\,M \qquad\qquad 1\,M$$
$$\Delta G^\circ = 4130 \quad \text{cal/mole}.$$

These conditions, as we shall see later are defined as *standard conditions*, and therefore the ΔG for this special situation is labeled ΔG° and called the *standard* free energy change. Turning to some other conditions, we might pick the concentrations:

$$\text{Ala(aq)} + \text{Gly(aq)} \rightleftharpoons \text{Ala-Gly(aq)} + \text{H}_2\text{O(l)},$$
$$0.1\,M \qquad 0.1\,M \qquad 1.25 \times 10^{-5}\,M$$
$$\Delta G = 0.$$

These concentrations happen to be one set at which this reaction is at equilibrium; therefore ΔG is zero. Another set of concentrations, neither all standard nor equilibrium values, is:

$$\text{Ala(aq)} + \text{Gly(aq)} \rightleftharpoons \text{Ala—Gly(aq)} + \text{H}_2\text{O(l)},$$
$$1\,M \qquad 1\,M \qquad\qquad 1 \times 10^{-4}\,M$$
$$\Delta G = -1350 \quad \text{cal/mole}.$$

The free energy change for this last set of concentrations, whose calculation depends on steps to be outlined in Chapter 5, is a negative number, neither ΔG° nor zero. From the three different values of ΔG, we see that depending on the concentrations of the species involved, the reaction of Eq. (3.7) may occur spontaneously as written, may be one at equilibrium, or may occur spontaneously in the reverse direction.

Thus ΔG depends on the states of the reactants and products, as well as on the nature of the transformation. To illustrate methods of computation, therefore, we shall divide our exposition into two sections. In the first section we consider only methods of calculating standard free energies ΔG°. Thereafter, we proceed to outline methods used to calculate ΔG accompanying changes in concentration or state and in this way cover a full range of conditions of biochemical interest.

4

Computations of Standard Free Energies

There are a number of general procedures for calculating $\Delta G°$. In any specific instance, the one chosen depends on the data available. A few typical examples will illustrate some useful procedures.

A. CALCULATIONS FROM EQUILIBRIUM CONSTANTS

If equilibrium data are available, the standard free energy change for the reaction may be computed directly from the relation

$$\Delta G° = -RT \ln K, \tag{4.1}$$

where R is the gas law constant (1.987 cal/mole/deg) and K the equilibrium constant. Equation (4.1) will be accepted without proof, but it can be derived readily once we know the relationship between free energy (or chemical potential) and concentration.*

As a specific numerical example of such a calculation, let us consider the following step in glycogen metabolism:

$$\text{Glucose 1-phosphate} \rightleftharpoons \text{Glucose 6-phosphate.} \tag{4.2}$$

This conversion is catalyzed by the enzyme phosphoglucomutase. The equilibrium has been studied extensively; at 25°C and pH 7, K is 17. Therefore,

$$\Delta G° = -(1.987)(298.15)(2.303)(\log_{10} 17)$$
$$= -1700 \quad \text{cal/mole.} \tag{4.3}$$

Thus, the transphosphorylation will occur spontaneously under *standard conditions*.

It might be appropriate at this juncture to emphasize an important and somewhat subtle point. Although we use equilibrium data to obtain the value of K inserted into Eq. (4.1) [or Eq. (4.3)], the free energy change computed is *not* for the transformation under equilibrium conditions. Quite the contrary, the ΔG obtained is $\Delta G°$, for *standard* conditions. Why this should be so is only apparent in an examination of the derivation of Eq. (4.1), which we have omitted. Nevertheless, it is a fact that from measurements on the equilibrium state, we can compute a free energy change for the *nonequilibrium* state. In this connection, we should recall that ΔG is zero for a reaction at equilibrium.

Let us also examine another specific example of a $\Delta G°$ calculation in which we penetrate a little further to the actual experimental measurements. A subject of long interest to protein biochemists is the thermodynamics of protein denaturation. One system that has been studied extensively in different ways is chymotrypsinogen. If its denaturation is reversible and can be represented by a two-state transition,

$$\text{N(native)} \rightleftharpoons \text{D(denatured)} \tag{4.4}$$

we can write for the equilibrium constant K

$$K = [\text{D}]/[\text{N}] \tag{4.5}$$

If this equilibrium constant could be evaluated, we could immediately compute $\Delta G°$ for denaturation from Eq. (4.1).

---

* See, for example, I. M. Klotz and R. M. Rosenberg, "Chemical Thermodynamics," 3rd Ed., pp. 136–138, 331–332. Benjamin, Menlo Park, California, 1972, or 4th Ed., 1986 (in press).

A. Calculations from Equilibrium Constants

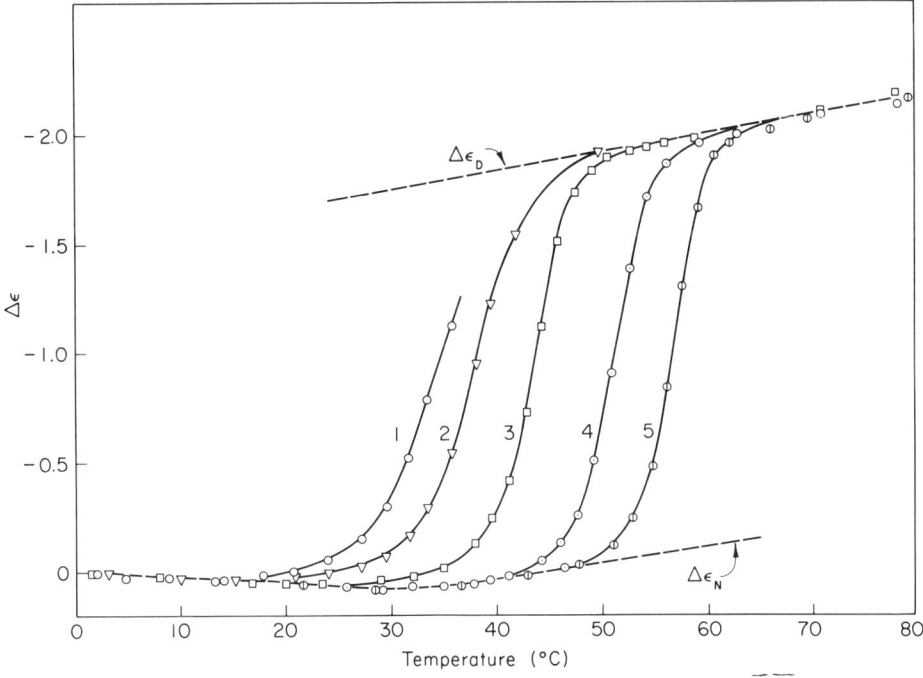

Fig. 4.1. Variation of $\Delta\varepsilon_{293\ nm}$ with temperature at pH values of 1.11 (○), 1.71 (▽), 2.07 (□), 2.56 (⊙) and 3.00 (⌽). Here $\Delta\varepsilon_N$ represents the difference extinction coefficient of the native protein and $\overline{\Delta\varepsilon}_D$ that for the denatured protein.

Proteins undergo changes in absorption of ultraviolet light when they are denatured. Such changes may be used, therefore, to follow the extent of denaturation. If heat is used to denature chymotrypsinogen at some specified pH, we find a sigmoid-type change in the ultraviolet difference spectrum* as is illustrated by any one of the curves in Fig. 4.1. The difference extinction coefficient $\Delta\varepsilon_N$ at 293 nm, for the fully native protein increases slightly with temperature, as does that for the fully denatured chymotrypsinogen $\Delta\varepsilon_D$. These small changes are shown in Fig. 4.1. Major changes in extinction occur when N → D. For example, curve 3 in Fig. 4.1 shows a sharp rise as the temperature goes above 40°C, as the native protein is converted more and more to the denatured form. The closer we approach the line for $\Delta\varepsilon_D$, the more nearly complete is the denaturation. At some intermediate temperature (for example, between 40° and 50°C in curve 3) we have a mixture of N and D, with an extinction coefficient $\Delta\varepsilon$. It is reasonable to assume that $\Delta\varepsilon - \Delta\varepsilon_N$, the distance

---

* J. F. Brandts, *J. Am. Chem. Soc.*, **86**, 4291 (1964); **89**, 4826 (1967).

upward to the denatured form should be proportional to the concentration of D, and that the remaining distance $\Delta\varepsilon_D - \Delta\varepsilon$ should be proportional to the concentration of N.* Thus, we may write

$$K = (\Delta\varepsilon - \Delta\varepsilon_N)/(\Delta\varepsilon_D - \Delta\varepsilon), \qquad (4.6)$$

and the equilibrium constant can be evaluated as a function of temperature, at each of a series of pH's.

At pH 1.7 and 43°C, $\Delta G°$ for denaturation of chymotrypsinogen is -1000 cal/mole; thus the denaturation will proceed spontaneously under standard concentration conditions. At the same pH and 34°C, $\Delta G°$ is $+1000$ cal/mole; thus the reverse reaction, *renaturation*, will occur spontaneously under standard conditions. Interestingly, in this reacting system, there are also some conditions, for example pH 1.7 and 38°C, for which $\Delta G°$ is zero, and, thus, native and denaturated forms would be in equilibrium with each other when they are present at standard state concentrations. In general, of course, the equilibrium state is *not* one in which the participants are also in the standard state.

The relationship between $\Delta G°$ and the equilibrium constant K [Eq. (4.1)] also leads us to recognize that relatively minor changes in structure of a biomacromolecule associated with small changes in free energy can have pronounced effects on biological function. For example, tetrameric hemoglobin, $\alpha_2\beta_2$, dissociates into pairs of dimers $\alpha\beta$ to different extents depending on the protein primary structure. The standard free energies of dissociation of normal adult hemoglobin A and of several mutants have been determined as shown in the following tabulation.

Hemoglobin	Substitution in mutant	$\Delta G°$ of dissociation (kcal/mole Hb)
A, normal	–	8.2
Richmond	102β; Asn → Thr	6.0
Kansas	102β; Asn → Thr	5.1
Georgia	95α; Pro → Leu	3.6

The difference in $\Delta G°$ between A and Richmond is only 2 kcal; but the mutant dissociation constant changes by 40 fold. Comparing A with Georgia we find a change in $\Delta G°$ of less than 5 kcal; the dissociation constant changes more than 2000 fold, and the behavior of this mutant oxygen carrier is altered markedly. This change in functional properties is produced by a single amino acid substitution, proline in the normal A being replaced by leucine. From a molecular viewpoint, we may be

---

* This assumption can be shown to be rigorously valid if the absorptions of the two species follow Beer's law.

A. Calculations from Equilibrium Constants

tempted to conclude that residue 95α is at the contact interface of two αβ dimers. We must recognize, however, that this is not necessarily so. The proline residue places constraints on the three-dimensional structure of a protein which are not present with a leucine residue and the effect on dissociation may be due to perturbations in structure at a distance from residue 95α. Structural arguments must be consistent with thermodynamic properties, but energetic quantities in themselves cannot establish the molecular origin of interactions or changes in interactions.

Among these considerations of equilibria, it is perhaps appropriate to draw attention to the thermodynamic basis of one of the fundamental principles of chemical transformations: an enzyme or catalyst cannot shift the equilibrium point. This statement is an automatic consequence of Eq. (4.1). If an enzyme were to shift an equilibrium in one direction, without itself undergoing any alteration in state, then K_{equil} would acquire a new value. Consequently $\Delta G°$ would differ in the presence of enzyme. However, $\Delta G°$ refers to the change in free energy under specific, standard conditions, the same conditions in the presence or absence of enzyme. Since standard conditions are the same whether catalyst is present or not, $\Delta G°$ can have only a single value. The chemical potential G, like other thermodynamic properties such as internal energy E or entropy S, depends only on the state of the system. In a change of state, ΔG depends only on the initial and final states, not on the physical or chemical path traversed to achieve this change. From these considerations it follows that $\Delta G°$ must have the same value in the presence or absence of a catalyst, and hence that an equilibrium constant cannot be shifted upon introduction of an enzyme.

In essence, what we are saying is that if an enzyme shifted an equilibrium, we would violate the laws of thermodynamics. A more impressive statement might be one that points out that if an enzyme affected an equilibrium, we could construct a perpetual motion machine. Consider for example a reaction

$$A + B \rightleftharpoons C + D + E$$

in which the volumes of the substances on the right are greater than those on the left. Let us assume for ease in visualization that A, B, C, D, and E are all gases. We allow these substances to reach equilibrium, in the absence of enzyme, in a vessel of the form shown in Fig. 4.2. The enzyme is held in a closed container at one corner of the large vessel. Once equilibrium is attained in the absence of enzyme, we slide the cover of the container over to a position which exposes the gases to the enzyme. If the enzyme affected the equilibrium (for example, shifted it to the right) the volume of the system would increase and the piston would be pushed

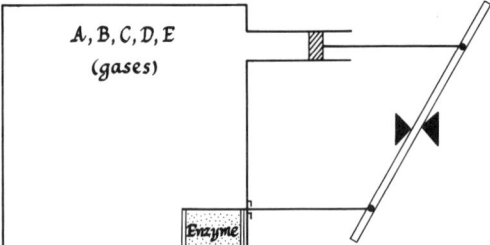

Fig. 4.2. Alleged perpetual motion machine.

out. By a simple arrangement of rods and levers attached to the piston, we could use a small part of the expansion force to slide back the cover of the enzyme container. The gases would then return to their initial equilibrium, the volume would decrease and the piston would move in.* As the piston moved in, the enzyme would become uncovered, and the cycle would be started again. This sequence of operations would recycle indefinitely. Thus, if an enzyme affected an equilibrium, we could construct a perpetual motion machine.

B. CALCULATIONS FROM OXIDATION–REDUCTION POTENTIALS

Tables of standard oxidation–reduction potentials, $\mathscr{E}°$, are available for many substances of biochemical interest.† The relationship for converting this information to $\Delta G°$ is

$$\Delta G° = -n\mathscr{F}\mathscr{E}°, \qquad (4.7)$$

where n is the number of moles of electrons transferred (per mole of transformation) and $\mathscr{F}$ is the Faraday constant (96,487 C/equiv, or 23,061 cal/V equiv).

Thus, if we ask whether ferrocytochrome c can reduce ferricytochrome f (at pH 7)

$$\text{Cyt } c \text{ Fe}^{2+} + \text{Cyt } f \text{ Fe}^{3+} \rightleftharpoons \text{Cyt } c \text{ Fe}^{3+} + \text{Cyt } f \text{ Fe}^{2+} \qquad (4.8)$$

we can obtain an answer if we can compute $\Delta G°$. To obtain the value of $\mathscr{E}°$, or $\mathscr{E}°'$ at pH 7, we find in Table 4.1 the oxidation–reduction potentials for

---

* The same net effect would be obtained if the enzyme shifted the equilibrium to the left.
† P. A. Loach, in "Handbook of Biochemistry," (H. A. Sober, ed.), 2nd Ed., pp. J-33–J-40. Chemical Rubber Co., Cleveland, 1970.

C. Calculations from Enthalpy and Entropy Changes

TABLE 4.1

Oxidation–Reduction Potentials at Room Temperature

System	$\mathscr{E}^{\circ\prime}$(V)	$\mathscr{E}^{\circ\prime}$(V) at pH 7
$\tfrac{1}{2}Cl_2 + e = Cl^-$	1.359	—
$\tfrac{1}{2}O_2 + 2e + 2H^+ = H_2O$	1.229	0.816
$Fe^{3+} + e = Fe^{2+}$	0.771	—
Cyt f $Fe^{3+} + e =$ Cyt f Fe^{2+}	—	0.365
$[Fe(CN)_6]^{3-} + e = [Fe(CN)_6]^{4-}$	—	0.36
Cyt c_5 $Fe^{3+} + e =$ Cyt c_5 Fe^{2+}	—	0.32
Cyt c $Fe^{3+} + e =$ Cyt c Fe^{2+}	—	0.254
Cyt b_2 $Fe^{3+} + e =$ Cyt b_2 Fe^{2+}	—	0.219
$HbFe^{3+} + e = HbFe^{2+}$	—	0.144
Fumarate $+ 2e + 2H^+ =$ Succinate	—	0.031
Rubredoxin $Fe^{3+} + e =$ Rubredoxin Fe^{2+}	—	−0.057
Pyruvate $+ 2e + 2H^+ =$ Lactate	—	−0.188
Acetaldehyde $+ 2e + 2H^+ =$ Ethanol	—	−0.197
$FAD + 2e + 2H^+ = FADH_2$	—	−0.219
$NAD^+ + 2e + H^+ = NADH$	−0.105	−0.320
Ferredoxin $Fe^{3+} + e =$ Ferredoxin Fe^{2+}	—	−0.413
$H^+ + e = \tfrac{1}{2}H_2$	0.000	−0.421

the half-reactions involving the participants in Eq. (4.8):

$$\text{Cyt } f \text{ Fe}^{3+} + e \rightleftharpoons \text{Cyt } f \text{ Fe}^{2+}, \quad \mathscr{E}^{\circ\prime} = 0.365 \text{ V},$$

$$\text{Cyt } c \text{ Fe}^{3+} + e \rightleftharpoons \text{Cyt } c \text{ Fe}^{2+}, \quad \mathscr{E}^{\circ\prime} = 0.254 \text{ V}.$$

If we subtract the second half-reaction from the first, then we obtain Eq. (4.8), for which the oxidation–reduction potential must be

$$\mathscr{E}^{\circ\prime} = 0.365 - 0.254 = 0.111 \quad \text{V}.$$

Therefore,

$$\Delta G^\circ = -(1)(23{,}061)(0.111)$$
$$= -2560 \quad \text{cal/mole}, \tag{4.9}$$

and the electron transfer can occur spontaneously.

C. CALCULATIONS FROM ENTHALPY AND ENTROPY CHANGES

A third approach to the calculation of free energy changes starts with the equation [see Eq. (3.4)]

$$\Delta G^\circ = \Delta H^\circ - T\,\Delta S^\circ. \tag{4.10}$$

If the enthalpy change $\Delta H°$ and the entropy change $\Delta S°$ for the reaction can be calculated, then $\Delta G°$ can be obtained promptly.

Generally, authoritative sources* list $\Delta Hf°$, the standard enthalpy of formation of a substance from its elements at the same temperature. It can be shown by simple algebra that for any reaction,

$$\Delta H° = \Delta Hf°_{products} - \Delta Hf°_{reactants}. \tag{4.11}$$

Similarly, authoritative tables frequently have values of $S°_T$, the absolute entropy of a substance at the temperature T. It is obvious that for any reaction

$$\Delta S° = S°_{products} - S°_{reactants}. \tag{4.12}$$

Let us apply these simple rules to a sample calculation. For each of the components of the reaction in the synthesis of (solid) leucylglycine we can find $\Delta Hf°$ and $S°$ (Table 4.2). These can be listed conveniently below each substance in the equation for the reaction

$$\text{DL-Leu (s) + Gly (s)} \rightleftharpoons \text{DL-Leu—Gly (s) + + H}_2\text{O (l)} \tag{4.13}$$

$\Delta Hf°$: -154.16 -128.4 -205.6 -68.317

$S°$: 49.5 24.7 67.2 16.716

where (s) represents the solid state and (l) the liquid. Thus, for this reaction

$$\Delta H° = -205{,}600 - 68{,}317 - (-154{,}160 - 128{,}460) = 8700 \quad \text{cal/mole},$$

$$\Delta S° = 67.2 + 16.716 - (49.5 + 24.7) = 8.7 \quad \text{cal/mole deg},$$

$$\Delta G° = 8700 - (298)(8.7) = 6100 \quad \text{cal/mole}.$$

Another example will illustrate the use of Eq. (4.10) together with a corollary principle of energetics. The hydrolysis of cytidine 2', 3'-cyclic phosphate has been studied widely, since this compound is a very convenient substrate for pancreatic ribonuclease. It would be desirable to know $\Delta G°$ for this hydrolysis. It is a relatively straightforward matter to

---

* *Selected Values of Chemical Thermodynamic Properties*, Nat. Bur. Standards (U.S.), Circ. 500, 1952; Technical Note 270-3, 1968; 270-4, 1969; 270-5, 1971; 270-6, 1971; 270-7, 1973; 270-8, 1981; J. O. Hutchens in "*Handbook of Biochemistry*," (H. A. Sober, ed.), 2nd Ed., pp. B-62–B-69. Chemical Rubber Co., Cleveland, 1970; "*Selected Values of Properties of Chemical Compounds*," Thermodynamics Research Center, Texas A&M University, College Station, Texas.

C. Calculations from Enthalphy and Entropy Changes

TABLE 4.2

Heats of Formation $\Delta Hf°$, and Absolute Entropies $S°$ at 25°C for Substances in the Solid State

Substance	$\Delta Hf°$ (kcal/mole)	$\Delta S°$ (cal/mole deg)
L-Alanine	−134.50	30.9
D-Arginine	—	59.9
L-Arginine · HCl	—	68.4
L-Asparagine	−188.7	41.7
L-Aspartic acid	−232.6	40.7
L-Cysteine	−126.7	40.6
L-Cystine	−249.6	67.1
L-Glutamic acid	−241.3	45.0
L-Glutamine	−197.5	46.6
Glycine	−128.4	24.7
L-Hydroxyproline	—	41.2
Hippuric acid	−147.71	57.1
L-Isoleucine	−152.5	49.7
L-Leucine	−152.4	50.6
D-Leucine	−152.4	50.6
DL-Leucine	−154.16	49.5
L-Lysine · HCl	—	63.2
L-Methionine	−181.2	55.3
L-Phenylalanine	−111.6	51.1
L-Proline	—	39.2
L-Serine	−173.6	35.7
L-Threonine	−192.9	36.5
L-Tryptophan	−99.2	60.0
L-Tyrosine	−160.5	51.2
L-Valine	−147.7	42.8
CO_2 (g)	−94.0518	51.08
H_2O (l)	−68.3174	16.716
DL-Alanylglycine	−186.0	51.0
DL-Leucylglycine	−205.6	67.2
DL-Leucylglycylglycine	−255.8	—

obtain calorimetric measurements directly for the reaction of interest:

$$\text{Cytidine 2',3'-cyclic phosphate (aq)} + H_2O\,(l) \xrightarrow{\Delta H°} \text{Cytidine 3'-phosphate (aq)}. \quad (4.14)$$

However, it is not feasible to obtain $\Delta S°$ directly for this reaction. On the other hand it is possible to evaluate $\Delta S°'$ for a series of steps that start with the same reactants and end with the same products as are shown in

Eq. (4.14) but which involve a different reaction pathway, i.e.,

$$
\begin{array}{ccc}
\text{Cytidine 2',3'-cyclic} & & \\
\text{phosphate (aq)} + H_2O\,(l) & & \text{Cytidine 3'-phosphate (aq)} \\
\Big\downarrow \Delta S_1^\circ & & \Big\uparrow \Delta S_3^\circ \\
\text{Cytidine 2',3'-cyclic} & \xrightarrow{\Delta S_2^\circ} & \text{Cytidine 3'-phosphate (s)} \\
\text{phosphate (s)} + H_2O\,(l) & &
\end{array}
\qquad (4.15)
$$

As mentioned earlier with regard to the free energy, the laws of thermodynamics also require that

$$\Delta S_1^\circ + \Delta S_2^\circ + \Delta S_3^\circ = \Delta S^\circ \quad \text{for Eq. (4.14),} \qquad (4.16)$$

because the pathway used in going from reactants to products is irrelevant insofar as values of the thermodynamic properties are concerned. Thus, it is possible to compute ΔG° for Eq. (4.14). Specifically, the values of the thermodynamic quantities at 25°C are*

$$
\begin{aligned}
\Delta S_1^\circ &= +8.22 \quad \text{cal/mole deg,} \\
\Delta S_2^\circ &= -9.9 \quad \text{cal/mole deg,} \\
\Delta S_3^\circ &= +8.28 \quad \text{cal/mole deg,} \\
\Delta S^\circ &= 6.6 \quad \text{cal/mole deg,} \\
\Delta H^\circ &= -2800 \quad \text{cal/mole,} \\
\Delta G^\circ &= -4800 \quad \text{cal/mole.}
\end{aligned}
\qquad (4.17)
$$

The negative value of ΔG° corresponds with the known fact that cytidine cyclic phosphate will hydrolyze spontaneously in the presence of the enzyme ribonuclease.

D. CALCULATIONS FROM FREE ENERGIES OF FORMATION

Thermodynamic data have now been accumulated for many substances involved in metabolic and biosynthetic processes. One concise way of summarizing this information is in terms of ΔGf°, the standard free energy of formation of the substance from the elements. A list of ΔGf° is presented in Table 4.3.

---

* J. T. Bahr, R. E. Cathou and G. G. Hammes, *J. Biol. Chem.*, 240, 3372 (1965).

TABLE 4.3

Standard Free Energies of Formation $Gf°$ at 25°C

	$\Delta Gf°$ (kcal/mole)	
Substance	Pure solid	In aqueous solution
Acetic acid (1)	−93.75	−95.48
Acetate ion	—	−88.99
cis-Aconitate^{3-}	—	−220.51
L-Alanine	−88.40	−88.75
DL-Alanine	—	−89.11
DL-Alanylglycine	—	−114.57
NH_3 (g)	−3.98	−6.37
NH_4^+ ion	—	−19.00
L-Aspartic acid	174.76	−172.31
L-Aspartate$^-$	—	−166.99
Carbon (graphite)	0.00	—
CO_2 (g)	−94.26	−92.31
HCO_3^-	—	−140.31
Citrate^{3-}	—	−279.24
Creatine	—	−63.17
Creatinine	—	−6.91
Cystine	−163.55	−159.00
Cysteine	−82.08	−81.21
Ethanol (1)	−41.77	−43.39
Fructose	—	−218.78
Fumaric acid	−156.49	−154.67
Fumarate^{2-}	—	−144.41
α-D-Galactose	−220.00	−220.73
α-D-Glucose	−217.56	−219.22
L-Glutamate$^-$	—	−165.87
Glycerol (1)	−114.02	−116.76
Glycine	−88.61	−89.14
Glycogen (per glucose unit)	—	−158.3
OH^- ion	—	−37.60
Hydrogen (g)	0.00	—
H^+ ion	—	0.00
α-Ketoglutarate^{2-}	—	−190.62
Lactate$^-$	—	−123.76
α-Lactose	—	−362.15
β-Lactose	—	−375.26
L-Leucine	−82.63	−81.68
DL-Leucine	—	−81.76
DL-Leucylglycine	—	−110.90
Malate^{2-}	—	−201.98
β-Maltose	—	−357.80
Nitrogen (g)	0.00	—
Oxalacetate^{2-}	—	−190.53
Oxygen (g)	0.00	—
Pyruvate$^-$	—	−113.44
Succinic acid	−178.68	−178.39
Succinate^{2-}	—	−164.97
Sucrose	−369.20	−370.90
Urea	−47.12	−48.72
Water (1)	—	56.69

We may illustrate the use of such a table with the following simple problem. Suppose we wish to find $\Delta G°$ for the racemization of L-alanine:

$$\text{L-Ala (aq)} \rightleftharpoons \text{DL-Ala (aq)}. \qquad (4.18)$$

From the listing in Table 4.3, it follows that for the formation of DL-Ala from its elements,

$$3\text{C (graphite)} + 3\tfrac{1}{2}\text{H}_2(g) + \text{O}_2(g) + \tfrac{1}{2}\text{N}(g) \rightleftharpoons \text{DL-Ala(aq)};$$

$$\Delta G° = \Delta Gf° \text{ (DL-Ala)} \qquad (4.19)$$
$$= -89{,}110 \text{ cal/mole}.$$

Correspondingly, for the reverse reaction for L-Ala,

$$\text{L-Ala (aq)} \rightleftharpoons 3\text{C (graphite)} + 3\tfrac{1}{2}\text{H}_2(g) + \text{O}_2(g) + \tfrac{1}{2}\text{N}_2(g);$$

$$\Delta G° = -\Delta Gf° \text{(L-Ala)} \qquad (4.20)$$
$$= +88{,}759 \text{ cal/mole}.$$

Addition of Eq. (4.19) and (4.20) leads to

$$\text{L-Ala (aq)} \rightleftharpoons \text{DL-Ala (aq)},$$
$$\Delta G° = -360 \text{ cal/mole}.$$

Thus, racemization could occur spontaneously; but thermodynamics says nothing about the rate of the reaction.

With this example we conclude our survey of methods for the computation of *standard* free energies. Now we must turn our attention to nonstandard conditions.

EXERCISES

1. Using the table of free energies of formation, find $\Delta G°$ for the reaction

$$\text{Glucose (aq)} + \text{O}_2(g) = 2\,\text{Pyruvate}^{-1}(aq) + 2\text{H}^+(aq) + 2\text{H}_2\text{O (l)}.$$

   Answer: -121.04 kcal/mole

2. Can the dehydration of malate to fumarate (in the Krebs cycle) occur spontaneously?
   Answer: $\Delta G° = 0.88$ kcal/mole.

3. The equilibrium constant for the reaction

$$\text{Ala (aq)} + \text{Gly (aq)} \rightleftharpoons \text{Ala—Gly (aq)} + \text{H}_2\text{O}$$

   is 1.25×10^{-3} at 38°C. Assuming that the equilibrium constant is essentially the same at 25°C and using a table of $\Delta Gf°$'s for L-alanine and glycine, find the standard free energy of formation of alanylglycine in aqueous solution.

Exercises

Answer: −117.68 kcal/mole.

4. For the conformational transition T → R in aspartate transcarbamoylase in the temperature range of 20–30°C,

$$\text{ATCase}^T \rightleftharpoons \text{ATCase}^R$$

the standard free energy change $\Delta G°$ in an aqueous solvent of defined composition has been determined to be 3.3 kcal/mole [G. J. Howlett, M. N. Blackburn, J. G. Compton, and H. K. Schachman, *Biochemistry* **126**, 5091–5099 (1977)]. Calorimetric measurements [A. Shrake, A. Ginsburg, and H. K. Schachman, *J. Biol. Chem.* **256**, 5005–5015 (1981)] lead to a $\Delta H°$ of −6 kcal/mole for the transition. What is the entropy change for T → R transition?

Answer: −31 cal/mole deg.

5. (a) In aqueous solution at pH 7 and 298 K, the standard free energies of formation (in kilocalories per mole) of leucine, glycine, the dipeptide, and H_2O can be found from Table 4.3. Calculate $\Delta G°$ at pH 7 and 298 K for the hydrolysis of the dipeptide by a peptidase enzyme.

Answer: −3.3 kcal/mole.

(b) In the cell *in vivo*, proteins are synthesized by activating amino acids by the following step:

$$\text{Amino acid} + \text{tRNA} + \text{ATP} \rightleftharpoons \text{Amino acyl-tRNA} + \text{AMP} + 2\text{P}$$

for which $\Delta G°$ at pH 7 and 298 K is −7.0 kcal/mole. For the hydrolysis of ATP to give 2 moles of inorganic P

$$\text{ATP} + H_2O \rightleftharpoons \text{AMP} + 2\text{P}$$

$\Delta G°$ is −14.6 kcal/mole. Using these data and the value you obtained for $\Delta G°$ of hydrolysis in (a), calculate $\Delta G°$ at pH 7 and 298 K for dipeptide biosynthesis *in vivo*:

$$_1\text{Amino acyl-tRNA} + {}_2\text{Amino acid} = {}_1\text{Amino acyl} - {}_2\text{Amino acid} + \text{tRNA}$$

Answer: −4.3 kcal/mole.

6. One of the metabolic pathways of pyruvic acid involves decarboxylation to acetaldehyde and carbon dioxide. For the conditions listed in the equation

$$\underset{\text{CH}_3\overset{\overset{\displaystyle O}{\|}}{\text{C}}\text{COOH (l)}}{} \rightleftharpoons \underset{\text{CH}_3\overset{\overset{\displaystyle O}{\|}}{\text{C}}\text{H (g)}}{} + CO_2 \text{ (g)}$$

the equilibrium constant at 25°C is $K = 1.79 \times 10^{11}$. The standard free energy of formation of acetaldehyde (g) is −31,860 cal/mole;

(a) Calculate $\Delta G°$ for the decomposition in the equation shown.

Answer: −15,370 cal/mole.

(b) What is $\Delta G_f°$ of pyruvic acid (1)?
Answer: $-110{,}750$ cal/mole.
7. The deamination of aspartic acid

$$^-OOC-CH_2-\underset{\underset{NH_3^+}{|}}{CH}-COO^- \rightleftharpoons {}^-OOC-CH=CH-COO^- + NH_4^+ \quad (1)$$

is a reversible reaction catalyzed by the enzyme aspartase. For DL-aspartic acid the equilibrium constant over a range of temperature [J. L. Bada and S. L. Miller, *Biochemistry* **7**, 3403 (1968)] may be expressed by the equation

$$\log K_{DL} = 8.188 - (2315.5/T) - 0.01025T. \quad (2)$$

(a) What is the value of $\Delta G°$ at 25°C?
Answer: 3.60 kcal/mole.
(b) Using the thermodynamic relation

$$\frac{d \ln K}{dT} = \frac{\Delta H°}{RT^2}$$

find an equation for $\Delta H°$ as a function of the temperature.
(c) Find the value of $\Delta H°$ at 25°C.
Answer: 6.42 kcal/mole.
(d) Find $\Delta S°$ at 25°C.
Answer: 9.50 cal/mole deg.
(e) Using the thermodynamic relation

$$\Delta C_p° = \left(\frac{\partial \Delta H°}{\partial T}\right)_P$$

find $\Delta C_p°$ at 25°C for the chemical transformation in Eq. (1).
Answer: -28 cal/mole deg.
(f) $\Delta G_f°$ of fumarate ion and ammonium ion are listed in Table 4.2. Find $\Delta G_f°$ of DL-aspartate ion.
8. In the normal course of visual events the *cis*-retinal in rhodopsin is converted to the all-trans geometric isomer:

$$\text{11-cis} \rightleftharpoons \text{all-trans}$$

In a nonpolar environment, the $\Delta S°$ for this conversion is 4.4 cal/mole deg and the $\Delta H°$ is 150 cal/mole [R. Hubbard, *J. Biol. Chem.* **241**, 1814 (1966)]. Find the equilibrium constant

$$K = \frac{\text{(all-trans)}}{\text{(11-cis)}}$$

Answer: 7.1

5

The Dependence of Chemical Potential on Concentration

Once we know $\Delta G°$, we could calculate ΔG for any other set of conditions if we knew how to obtain the change in free energy as we change the state of each participant in the reaction. The solution of the last step becomes straightforward once we set forth the fundamental relationship between chemical potential (free energy) and concentration.

A. FUNDAMENTAL RELATIONSHIP

The basic statement relates free energy G to a "generalized concentration" a which is actually called the *activity*:

$$G = RT \ln a + \text{constant}. \qquad (5.1)$$

We shall discuss, in a moment, the manner in which activity is reduced to

TABLE 5.1

Activity Relations

State of substance	Nature of activity
Gas	
Ideal	P (pressure, in atmospheres)
Real	f (fugacity) $= P\gamma$
Pure solid	$a = 1$
Pure liquid	$a = 1$
Solutions	
Nonelectrolytes	
Solvent	$a_1 = N_1\gamma_1$
Solute (practical standard state)	$a_2 = m\gamma_2$
Solute (unitary or rational standard state)	$a_2 = N_2\gamma_2$
Electrolytes	
Solvent	$a_1 = N_1\gamma_1$
Solute (e.g., NaCl)	$a_2 = m^2\gamma_2^2$

experimentally measured concentrations. At this point let us first apply Eq. (5.1) to the problem of calculating ΔG for the process of changing the state of any substance A from one condition to another:

$$A\,(\text{conc}) \rightleftharpoons A\,(\text{conc}'). \tag{5.2}$$

It follows from Eq. (5.1) that ΔG for the reaction in Eq. (5.2) is given by

$$\Delta G = G' - G = RT \ln a' - RT \ln a = RT \ln(a'/a) \tag{5.3}$$

To make actual computations, we now need to specify the relationships between a and experimental measurements. These are summarized in Table 5.1.

For a gas, a may usually be replaced by the pressure P or by partial pressure in a mixture. For most biological problems, the gas pressure does not exceed 1 atm, and for most gases, ideal behavior may be assumed at such low pressures. At higher pressures, it may be necessary to use a corrected pressure $P\gamma$, where γ is the correction factor necessary to give correct answers in free energy calculations.

For a pure solid or a pure liquid, a is defined as 1. This is merely another way of saying that the pure solid or pure liquid is the standard state for that substance.

Our most common calculations will involve substances in solution. For the solvent, we always define the activity in terms of its mole fraction N_1 and the correction factor γ_1, the activity coefficient; thus $a_1 = N_1\gamma_1$. In many solutions, the amount of dissolved solute, even if large in weight, does not contribute much to the total moles present, and hence N_1 is frequently near unity. Likewise γ_1 is often near unity except in solutions containing high concentrations of solute.

When we turn to the solute we must distinguish between non-electrolytes and electrolytes. In both types, the common or practical unit of concentration is molality m. The activity of a nonelectrolyte a_2 is given by the product $m\gamma_2$. For an electrolyte, the result is more complicated, depending on the number of ions produced. For a uni-univalent electrolyte, e.g. NaCl, $a_2 = m^2\gamma_2^2$. For solutes in very dilute solution, γ_2 approaches unity, and hence a_2 can be replaced by m (for nonelectrolytes) or by m^2 (for uni-univalent electrolytes) alone.

Finally, we should note that an alternative standard state, the unitary (or rational) standard state, has been used with some frequency recently in biochemical calculations. With this choice of standard state, we obtain "unitary free energies."* In essence, this choice measures solute concentration in mole fraction units, N_2, and a_2 becomes $N_2\gamma_2$.

Having defined the activity a in terms of experimental quantities, we should turn back briefly to a more precise specification of the meaning of $\Delta G°$, the standard free energy change. We now make the general and precise statement that $\Delta G°$ is the free energy change for a reaction in which each of the participants is at an *activity of 1*. When it has an activity of unity, the substance is in its standard state. For an ideal gas, this means the gas at 1 atm pressure. For a solute, this means when $a_2 = 1$. In the practical choice, the activity is unity somewhere near a molality of 1. If γ_2 is unimportant (i.e., essentially unity), then the activity is truly unity at unit molality, i.e., the standard state of the solute is the one-molal solution. In many problems the available data are not accurate enough to evaluate γ_2, and hence, to a first approximation, we may set $a_2 = m$. If we choose the less common "unitary" standard state, then concentrations of solute are measured in mole fraction N_2, and the activity is unity when $N_2\gamma_2$ is unity.

B. ILLUSTRATIVE CALCULATIONS

1. Free Energy of Dilution

Thus, in principle, we can compute ΔG for any concentrations, if we know ΔG for any single set of specifications at the same temperature. Such computations are essential in any transposition of free energy data from *in vitro* experiments to reasoning about behavior *in vivo*, for in physiological

---

* R. W. Gurney, "Ionic Processes in Solution," p. 90. McGraw-Hill, New York, 1953; W. Kauzmann, *Adv. Protein Chem.* **14**, 1 (1959).

media concentrations of the participating substances may differ by orders of magnitude from those convenient for direct experimental measurement. Thus, ΔG for a reaction occurring at physiological concentrations may be markedly different from $\Delta G°$, the change in free energy for standard conditions.

It is a simple matter to set up a general relationship between ΔG for any set of conditions and $\Delta G°$ which is normally provided in reference sources. We need merely add a suitable set of equations that express the necessary dilution steps and compute ΔG for each step by application of Eq. (5.1). For example, for the hydrolysis of adenosine triphosphate (ATP) to adenosine diphosphate (ADP) under standard* conditions, which we may represent by

$$\text{ATP}^{4-}\ (1\ M) + \text{H}_2\text{O} \rightleftharpoons \text{ADP}^{2-}\ (1\ M) + \text{HPO}_4^{2-}\ (1\ M),$$

$$\Delta G° = -7000 \quad \text{cal/mole}, \tag{5.4}$$

the change in chemical potential is near -7 kcal/mole.[†] We should like to have ΔG for concentrations different from $1\ M$, e.g., physiological concentrations. To obtain this value let us add the following three equations to Eq. (5.4):

$$\text{ATP}^{4-}\ (\text{physiol conc}) \rightleftharpoons \text{ATP}^{4-}\ (1\ \text{M}),$$

$$\Delta G = RT\ \ln(1/C_{\text{ATP(physiol conc)}}); \tag{5.5}$$

$$\text{ADP}^{2-}\ (1\ M) \rightleftharpoons \text{ADP}^{2-}\ (\text{physiol conc}),$$

$$\Delta G = RT\ \ln(C_{\text{ADP(physiol conc)}}/1); \tag{5.6}$$

$$\text{HPO}_4^{2-}\ (1\ M) \rightleftharpoons \text{HPO}_4^{2-}\ (\text{physiol conc}),$$

$$\Delta G = RT\ \ln(C_{\text{HPO}_4\text{(physiol conc)}}/1). \tag{5.7}$$

The result of adding these chemical equations is

$$\text{ATP}^{4-}\ (\text{physiol conc}) + \text{H}_2\text{O} \rightleftharpoons \text{ADP}^{2-}\ (\text{physiol conc}) + \text{HPO}_4^{2-}\ (\text{physiol conc}), \tag{5.8}$$

which differs from Eq. (5.4) only in the concentrations of the participants. Since the chemical reaction [Eq. (5.8)] is obtained by addition of the chemical reactions in Eqs. (5.4)–(5.7), ΔG for Eq. (5.8) is obtainable by

---

* For this problem, we assume γ's are unity, and that concentrations of solute may be expressed in molarities.

[†] M. F. Morales, J. Botts, J. J. Blum, and T. L. Hill, *Physiol. Rev.* **35**, 475 (1955).

B. Illustrative Calculations

summation of the ΔG's for Eqs. (5.4)–(5.7). Hence we may write

$$\Delta G_{\text{physiol}} = \Delta G° + RT \ln(1/C_{\text{ATP(physiol)}})$$
$$+ RT \ln C_{\text{ADP(physiol conc)}}$$
$$+ RT \ln C_{\text{HPO}_4\text{(physiol conc)}}$$
$$= \Delta G° + RT \ln(C_{\text{ADP}} C_{\text{HPO}_4}/C_{\text{ATP}})_{\text{physiol conc}} \quad (5.9)$$

Numerical values for $\Delta G_{\text{physiol}}$ can be obtained only when the physiological concentrations of ATP, ADP, and HPO_4^{2-} can be estimated. Even without these concentrations, however, we can readily see that under physiological conditions ΔG is likely to be appreciably more negative than $\Delta G°$, which is -7 kcal/mole. As a rough guess we could assume that the concentrations of ATP and ADP would be of the same order of magnitude and thus would cancel inside the logarithm in Eq. (5.9). Hence, the HPO_4^{2-} concentration, being a small decimal number, would make $RT \ln C_{\text{HPO}_4}$ a large negative number, to be added to -7 kcal/mole for $\Delta G°$. Thus, ΔG for the hydrolysis of ATP at physiological concentrations would be appreciably more negative than -7 kcal/mole.

2. $\Delta Gf°$ for Dissolved Substances

As a sample problem let us consider a calculation of the standard free energy of formation of succinic acid, $\text{HOOC–CH}_2\text{–CH}_2\text{–COOH}$, in aqueous solution at 25°C, assuming that we know $\Delta Gf°$ of solid succinic acid (see Table 4.3). To the equation for the formation of solid $(\text{CH}_2\text{COOH})_2$ from its elements,

$$4\text{C (graphite)} + 3\text{H}_2 \text{(g)} + 2\text{O}_2 \text{(g)} \rightleftharpoons \begin{array}{c} \text{CH}_2\text{COOH} \\ | \\ \text{CH}_2\text{COOH} \end{array} \text{(s)}, \quad (5.10)$$

$$\Delta Gf° = -178{,}680 \quad \text{cal/mole},$$

we add an equation for bringing the solid into a saturated aqueous solution,

$$\begin{array}{c} \text{CH}_2\text{COOH} \\ | \\ \text{CH}_2\text{COOH} \end{array} \text{(s)} \rightleftharpoons \begin{array}{c} \text{CH}_2\text{COOH} \\ | \\ \text{CH}_2\text{COOH} \end{array} \text{(aq, sat)}, \quad (5.11)$$

$$\Delta G = 0.$$

For the reaction in Eq. (5.11) ΔG is zero since this is a transfer at equilibrium. Now we add further the equation

$$\begin{array}{c} \text{CH}_2\text{COOH} \\ | \\ \text{CH}_2\text{COOH} \end{array} \text{(aq, sat)} \rightleftharpoons \begin{array}{c} \text{CH}_2\text{COOH} \\ | \\ \text{CH}_2\text{COOH} \end{array} \text{(aq, } a_2 = 1\text{),} \qquad (5.12)$$

$$\Delta G = ?$$

For this transfer

$$\Delta G = RT \ln(a_{\text{right}}/a_{\text{left}}). \qquad (5.13)$$

The activity on the right-hand side of Eq. (5.12) is unity by choice. For a_{left} we must keep in mind that we are dealing with *nonionized* succinic acid in the saturated aqueous solution at 25°C. Hence:

$$a_{\text{left}} = m_{\substack{\text{nonionized} \\ \text{in sat soln}}} \gamma = m_{\substack{\text{total} \\ \text{in sat soln}}} (1 - \text{fraction ionized}) \gamma \qquad (5.14)$$

The total concentration in the saturated solution, m_{total}, is easily measured experimentally. From the known pK_a of succinic acid in aqueous solution, or an appropriate pH measurement, we can compute the fraction ionized. The activity coefficient must be obtained from suitable measurements of the colligative properties of this solution. All of this information has been accumulated experimentally. Hence, we may write:

$$a_{\text{left}} = (0.715)(1 - 0.0112)(0.87) \qquad (5.15)$$

Inserting this information into Eq. (5.13) we find:

$$\Delta G = 290 \quad \text{cal/mole,} \qquad (5.16)$$

and this is the quantity to be used for the free energy change of reaction (5.12). Thus, if we add Eqs. (5.10)–(5.12) we obtain:

$$4\text{C (graphite)} + 3\text{H}_2 \text{ (g)} + 2\text{O}_2 \text{ (g)} \rightleftharpoons \begin{array}{c} \text{CH}_2\text{COOH} \\ | \\ \text{CH}_2\text{COOH} \end{array} \text{(aq),} \qquad (5.17)$$

$$\Delta G_f^\circ = -178{,}390 \quad \text{cal/mole}$$

This is the way in which the value listed in Table 4.3 was obtained.

We can extend this problem by now asking for ΔG_f° of monosuccinate ion, $(\text{CH}_2\text{COOH})(\text{CH}_2\text{COO}^-)$, in solution. This free energy of formation is obtained by adding the following equation to Eq. (5.17):

$$\begin{array}{c} \text{CH}_2\text{COOH} \\ | \\ \text{CH}_2\text{COOH} \end{array} \text{(aq, } a_2=1\text{)} \rightleftharpoons \text{H}^+ \text{(aq, } a_2=1\text{)} + \begin{array}{c} \text{CH}_2\text{—COO}^- \\ | \\ \text{CH}_2\text{COOH} \end{array} \text{(aq, } a_2=1\text{),} \qquad (5.18)$$

$$\Delta G^\circ = ?$$

Since every participant in this reaction is in its standard state, one

B. Illustrative Calculations

procedure for calculating the corresponding $\Delta G°$ merely makes use of [see Eq. (4.1)],

$$\Delta G° = -RT \ln K = +1365 \, pK_{a_1} = 5700 \quad \text{cal/mole}, \quad (5.19)$$

where K_{a_1} is the first acidity constant of succinic acid, $10^{-4.184}$. Addition of Eqs. (5.17) and (5.18) then gives

$$4C + 3H_2 + 2O_2 \rightleftharpoons H^+ \text{(aq)} + \begin{array}{c} CH_2COO^- \\ | \\ CH_2COOH \end{array} \text{(aq)}, \quad (5.20)$$

$$\Delta Gf° = -172{,}690 \quad \text{cal/mole}.$$

This is as far as we can proceed experimentally. For concise representation in tables, however, we agree to the following convention:

$$H^+ (\text{aq}, a_2 = 1) \rightleftharpoons \tfrac{1}{2} H_2 (\text{g}, a = 1);$$
$$\Delta G° = -\Delta Gf° \equiv 0. \quad (5.21)$$

In essence we arbitrarily define the standard free energy of formation of the H^+ ion as zero, in analogy to the corresponding convention with regard to $\mathscr{E}°$ for the H_2/H^+ electrode. Thus, if we add Eq. (5.21) to Eq. (5.20) we obtain $\Delta Gf°$ of monosuccinate ion:

$$4C + 2\tfrac{1}{2}H_2 + 2O_2 \rightleftharpoons \begin{array}{c} CH_2COO^- \\ | \\ CH_2COOH \end{array} \text{(aq)}, \quad (5.22)$$

$$\Delta Gf° = -172{,}690 \quad \text{cal/mole}.$$

We must recognize, however, that the actual $\Delta Gf°$ of a single ion in solution has not been (and, in fact, cannot be) determined. Values listed in Table 4.3 for ions are all relative values, relative to a conventional reference point of zero for H^+ ion.

Proceeding further with succinic acid we could obtain $\Delta Gf°$ for the dianion by adding:

$$\begin{array}{c} CH_2COO^- \\ | \\ CH_2COOH \end{array} (\text{aq}, a_2 = 1) \rightleftharpoons H^+ (\text{aq}, a_2 = 1) + \begin{array}{c} CH_2COO^- \\ | \\ CH_2COO^- \end{array} (\text{aq}, a_1 = 1) \quad (5.23)$$

$$\Delta G° = 1365 \, pK_{a_2}$$

to Eq. (5.20). Thereby we obtain:

$$4C + 3H_2 + 2O_2 \rightleftharpoons 2H^+ (\text{aq}) + (CH_2COO^-)_2 (\text{aq}), \quad (5.24)$$

$$\Delta Gf° = -164{,}970 \quad \text{cal/mole}.$$

Again, if we recognize the convention of Eq. (5.21), the $\Delta Gf°$ of succinate^{2-} is automatically $-164{,}970$ cal/mole (see Table 4.3).

Finally, in connection with the succinate problem, we should turn to one further consideration. Biochemical reactions are of primary interest at physiological pH. It is convenient, therefore, to know thermodynamic quantities at pH 7. To find the free energy of formation of succinate^{2-}, for example, at pH 7, we add to Eq. (5.24) the following equation:

$$2H^+ (aq, a_2 = 1) \rightleftharpoons 2H^+ (a_2 = 10^{-7}),$$
$$\Delta G = 2RT \ln(10^{-7}/1) = -19{,}100 \quad \text{cal.} \tag{5.25}$$

Thus, we obtain:

$$4C + 3H_2 + 2O_2 = 2H^+ (a_2 = 10^{-7}) + (CH_2COO^-)_2 (aq, a_2 = 1),$$
$$\Delta Gf \equiv \Delta Gf^{\circ\prime} = -184{,}070 \quad \text{cal/mole.} \tag{5.26}$$

Since every participant in reaction (5.26) except the H^+ ion is in its standard state, we agree to call the ΔGf a $\Delta Gf^{\circ\prime}$, the prime emphasizing that the H^+ is not at unit activity. Here too we can agree further that:

$$\tfrac{1}{2} H_2 = H^+ (a_2 = 10^{-7}),$$
$$\Delta Gf^{\circ\prime} = 0 \tag{5.27}$$

In this case then $\Delta Gf^{\circ\prime}$ of succinate^{-2} ion at pH 7 is automatically -184.07 kcal/mole.

Let us consider another example of a ΔG° calculation for a dissolved substance, one in which a gas is involved. For the first step in the oxygenation of hemoglobin at 19°C,

$$Hb(aq) + O_2(g) = HbO_2(aq),$$
$$\Delta G^\circ = -2590 \quad \text{cal/mole.} \tag{5.28}$$

The ΔG° was computed from the known value of 85.5 atm^{-1} for the equilibrium constant for this reaction at 19°C. Let us now try to find ΔG° for the corresponding reaction with O_2 (aq, $a_2 = 1$) instead of $O_2(g)$.

For this calculation we make use of the known solubilities of oxygen in water. At 19°C, in equilibrium with air (in which the partial pressure of oxygen is 0.2 atm) the solubility of oxygen in water is 0.00023 m.

Let us consider, therefore, the following three transfers of oxygen, for each of which we can readily calculate ΔG:

$$O_2 (g, P = 1) \rightleftharpoons O_2 (g, P = 0.2),$$
$$\Delta G = RT \ln(0.2/1) = -940 \quad \text{cal/mole,} \tag{5.29}$$

$$O_2 (g, P = 0.2) \rightleftharpoons O_2 (aq, sat),$$
$$\Delta G = 0 \,(\text{equil}), \tag{5.30}$$

B. Illustrative Calculations

$$O_2(aq, sat) \rightleftharpoons O_2(aq, a_2 = 1),$$
$$\Delta G = RT \ln(1/0.00023) = 4870 \quad \text{cal/mole}.$$
(5.31)

The net result of the addition of these three equation is:

$$O_2(g) = O_2(aq),$$
$$\Delta G° = 3930 \quad \text{cal/mole}.$$
(5.32)

Subtracting this equation from Eq. (5.28) we obtain,

$$Hb(aq) + O_2(aq) = HbO_2(aq),$$
$$\Delta G° = -6520 \quad \text{cal/mole}.$$
(5.33)

3. Free Energy of Hydrophobic Bonds

In essence, the stability of the hydrophobic bond is attributed to the tendency of an apolar group to remain in an apolar environment rather than to go out into a surrounding aqueous environment. Thus, one method of evaluating the stability of this bond is to determine the free energy of transfer $\Delta Gt°$ of an apolar group from an organic environment to water. A number of such calculations have been made and lead to positive values of $\Delta Gt°$ for apolar groups.

Since hydrophobic bonds are of special interest in protein molecules, it would be particularly desirable to evaluate $\Delta Gt°$ for apolar amino-acid side chains. However, these side chains cannot be plucked off by themselves. Nevertheless, a reasonable assumption is that:

$$\Delta Gt°[R-CH(NH_3^+)COO^-] - \Delta Gt°[H-CH(NH_3^+)COO^-]$$
$$= \Delta Gt°(R \text{ group}), \quad (5.34)$$

that is, the free energy of transfer of the amino acid $R-CH(NH_3^+)COO^-$ is composed of two parts, that of the apolar R group, and that of the zwitterion base which may be approximated by glycine, $H-CH(NH_3^+)COO^-$.

Since amino acids (particularly glycine) are not very soluble in very nonpolar solvents, such as benzene or hexane, we may settle for alcohol as the prototype nonpolar solvent.* Thus, we have reduced our problem to the measurement of $\Delta Gt°$ of various amino acids as they are transferred from alcohol to water. Since the procedure is the same for all the amino acids, we shall consider a detailed calculation for only one, glycine.

The transfer process cannot be conveniently carried out directly, at least not for thermodynamic computations. A series of steps that can make

---

* It is also feasible to use transfer to the gas phase as a model. See R. Wolfenden, L. Andersson, P. M. Cullis, and C. C. B. Southgate, *Biochemistry* **20**, 849 (1981).

use of the most readily accessible experimental data are the following:

$$\text{Gly (alcohol, sat soln)} \rightleftharpoons \text{Gly (s)};$$
$$\Delta G = 0 \tag{5.35}$$

$$\text{Gly (alcohol, } a_2 = 1) \rightleftharpoons \text{Gly (alcohol, sat soln)}$$
$$\Delta G = RT \ln(N_{\text{alc sat}} \, \gamma_{\text{alc sat}}/1) \tag{5.36}$$

$$\text{Gly (s)} \rightleftharpoons \text{Gly (aq, sat soln)};$$
$$\Delta G = 0 \tag{5.37}$$

$$\text{Gly(aq, sat soln)} \rightleftharpoons \text{Gly(aq, } a_2 = 1);$$
$$\Delta G \rightleftharpoons RT \ln(1/N_{\text{aq sat}} \gamma_{\text{aq sat}}) \tag{5.38}$$

For Eqs. (5.35) and (5.37) ΔG is zero since each of these is an equilibrium transfer. For the transfer in Eq. (5.36) ΔG is calculated by applying Eq. (5.3) using the unitary (or rational) standard state in which a of the solute is replaced by its mole fraction. The same procedure is used in Eq. (5.38). Thus, what we need to find are the solubilities of glycine in alcohol and in water ($N_{\text{alc sat}}$ and $N_{\text{aq sat}}$ respectively) and the corresponding activity coefficients. These data have been accumulated from the literature.*

If we now add Eqs. (5.35)–(5.38), we obtain

$$\text{Gly (alc, } a_2 = 1) \rightleftharpoons \text{Gly (aq, } a_2 = 1),$$

$$\Delta G = \Delta G t^\circ = RT \ln \frac{N_{\text{alc,sat}}}{N_{\text{aq,sat}}} + RT \ln \frac{\gamma_{\text{alc sat}}}{\gamma_{\text{aq,sat}}}. \tag{5.39}$$

In actual computations, the γ's must often be approximated. However, their contribution to $\Delta G t^\circ$ is always small compared to that of the N_{sat}'s.

From the solubilities alone, the values listed in Table 5.2 have been calculated for glycine and a series of amino acids with apolar side chains. For all of these, the free energy of transfer from alcohol to water is negative; that is, the transfer to an aqueous environment would occur spontaneously. However this $\Delta G t^\circ$ is composed of a contribution from the highly charged dipolar ion, $CH(NH_3^+)COO^-$, as well as from the apolar R group. If in each case we take the difference ΔG° as directed by Eq. (5.34) we obtain the values listed in the last column of Table 5.2. Here we have a measure of the stability of the hydrophobic side chain R. For isoleucine (Ile), for example, an input of 2970 cal/mole of free energy would be necessary to move the R group from the apolar environment to water.

---

* C. Tanford, *J. Am. Chem. Soc.* **84**, 4240 (1962); Y. Nozaki and C. Tanford, *J. Biol. Chem.* **238**, 4074 (1963).

TABLE 5.2

Unitary Free Energies of Transfer from Alcohol to Water at 25°C

Amino acid	$\Delta Gt°$	$\Delta G°$ (R group)
Gly	−4630	(0)
Ala	−3900	730
Val	−2940	1690
Leu	−2210	2420
Ile	−1690	2970
Phe	−1980	2650
Pro	−2060	2600

Correspondingly large free energies are necessary to move the other bulky apolar side chains. These numbers provide one basis for attempts to account for protein stability in terms of hydrophobic bonding.

EXERCISES

1. The hydrolysis of hippuric acid,

$$\text{Benzoylglycine} + H_2O \rightleftharpoons \text{benzoic acid} + \text{glycine}, \tag{1}$$

may be examined thermodynamically under a variety of conditions. Data necessary for some thermodynamic computations may be obtained from Table IV in the article by F. H. Carpenter [*J. Am. Chem. Soc.* **82**, 1111 (1960)].

(a) Consider first the situation in which all the components in Eq. (1) (at 298 K) except water are in the solid state, H_2O being liquid.

(1) Using data for $\Delta Hf°$ find $\Delta H°$ for reaction (1).
Answer: −3.94 kcal/mole.
(2) Using $S°$ data find $\Delta S°$ for reaction (1).
Answer: 7.82 cal/mole deg.
(3) Compute $\Delta G°$ for hydrolysis under these conditions.
Answer: −1611 cal/mole.

(b) To compute $\Delta G°$ (at 298 K) for reaction (1) with all substances dissolved in aqueous solution, we must find ΔG for transferring each substance from the solid state to solution.

(1) The solubility of benzoic acid in water is 0.028 m. Its pK_a is 4.20. The activity coefficient γ of the nonionized form in saturated aqueous solution is 1. Find $\Delta Gf°$ of nonionized benzoic acid in aqueous solution.
Answer: −56.45 kcal/mole.

(2) Find $\Delta G f°$ of H^+ + benzoate ion in aqueous solution.
Answer: -50.73 kcal/mole.
(3) Starting with the $\Delta G f°$ of (dissolved) nonionized hippuric acid, find $\Delta G f°$ of H^+ + hippurate ion in aqueous solution.
Answer: -81.19 kcal/mole.
(4) Compute $\Delta G°$ for reaction (1) with each component in solution at pH 6–7 (i.e., glycine isoelectric, hippuric and benzoic acids as the anions).
Answer: -1950 cal/mole

2. At an oxygen gas pressure of 0.13 atm and 25°C, the oxygen carrier hemocyanin (from squid) is 33% "saturated" with oxygen.
(a) Calculate the equilibrium constant for the reaction:

$$\text{Hcy (aq)} + O_2 \text{(g)} \rightleftharpoons \text{Hcy} \cdot O_2 \text{(aq)}. \quad (2)$$

Answer: 3.79 (atm^{-1}).
(b) Calculate $\Delta G°$ for the reaction

$$\text{Hcy (aq)} + O_2 \text{(aq)} \rightleftharpoons \text{Hcy} \cdot O_2 \text{(aq)}. \quad (3)$$

The solubility of pure oxygen in water at 1 atm and 25°C is 0.00117 moles/kg H_2O.
Answer: -4785 cal/mole

3. The equilibrium constants for the oxygenation reaction

$$O_2 \text{(g)} + \text{Hr (aq)} \rightleftharpoons \text{HrO}_2 \text{(aq)}, \quad (4)$$

where Hr represents hemerythrin (the oxygen-carrying pigment of *Phascolopsis gouldii*) are:

t (°C)	K (atm^{-1})
0	9120
25	380

(a) Using the thermodynamic relation $d \ln K/dT = \Delta H°/RT^2$ find $\Delta H°$, the heat of oxygenation.
Answer: -20.5 kcal/mole.
(b) Find $\Delta S°$, the entropy change for reaction (4) at 25°C.
Answer: 57.1 cal/mole deg.
(c) Find $\Delta G°$ for the reaction:

$$O_2 \text{(aq)} + \text{Hr (aq)} = \text{HrO}_2 \text{(aq)} \quad (5)$$

Answer: -7590 cal/mole.

4. The standard free energy of formation of L-alanine (s) is listed in Table 4.3. The solubility of L-Ala in water at 25°C is 1.87 m, and its

activity coefficient in a saturated solution is 1.05. Its acidity constants are: $pK_1 = 2.34$; $pK_2 = 9.69$.
 (a) Find $\Delta Gf°$ for L-Ala (aq). Compare your answer with that in Table 4.3.
 (b) Find $\Delta Gf°$ of L-Ala$^-$.
 (c) Find $\Delta Gf°$ of L-Ala$^+$.
 Answer: -91.99 kcal/mole.

5. The solubility of phenylalanine in water at 25° is 0.170 moles/kg solvent and 0.263 moles/kg in aqueous 6 M urea. Find the standard free energy of transfer of phenylalanine from water to aqueous urea solution using molality as the concentration unit and neglecting γ's.
 Answer: -260 cal/mole.
 The corresponding $\Delta Gt°$ for glycine is near zero. Does urea strengthen or weaken hydrophobic bonds?

6. From x-ray crystallographic diffraction, it has become apparent that a very large fraction of apolar side chains of proteins are actually on the surface of the macromolecule, exposed to solvent, rather than buried inside [I. M. Klotz, *Arch. Biochem. Biophys.* **138**, 704 (1970)]. For papain, for example, the moles of buried and accessible residues of alanine, valine, leucine, isoleucine, phenylalanine and typtophan are

	Buried	Exposed
Ala	6	8
Val	8	10
Leu	6	5
Ile	7	5
Phe	2	2
Trp	2	3

Assuming that this distribution in each case is an equilibrium one, calculate $\Delta G°$ of transfer of each hydrophobic side chain from the interior of the protein to the solvent.
 For a much more extensive assembly of such side chain distributions in a large number of proteins, see J. Janin, *Nature* **277**, 491 (1979).

6

Group-Transfer Potential: High-Energy Bond

One of the most useful concepts in quantitative biology, particularly in correlating the biochemical changes in intermediary metabolism, is that of the "high-energy bond." There are certain valid objections to the name, however, especially that it gives a misleading impression as to the nature of the quantity being considered. Since fundamentally, as we shall see shortly, we are talking about changes in chemical potential, ΔG, when certain groups are transferred from one molecule to another, a more appropriate name might be *group-transfer potential*.

A. COMPARISON OF TRANSFER POTENTIALS

There are, in fact, a number of other *transfer potentials* that are used in biochemical energetics, for example, acidity measured in terms of pK_a and oxidation–reduction potentials measured in terms of $\mathscr{E}°$. All of these

A. Comparison of Transfer Potentials

TABLE 6.1

Comparison of Types of Transfer Potential

	Proton-transfer potential (acidity)	Electron-transfer potential (redox potential)	Group-transfer potential (high-energy bond)
Concise equation	$AH \rightleftharpoons A^- + H^+$	$A \rightleftharpoons A^+ + e$	$A \sim PO_4 \rightleftharpoons A + PO_4$
Equation with acceptor	$AH + H_2O \rightleftharpoons A^- + H_3O^+$	$A + H^+ \rightleftharpoons A^+ + \frac{1}{2}H_2$	$A \sim PO_4 + H_2O \rightleftharpoons A\text{–}OH + HPO_4$
Measure of transfer potential	$pK_a = \dfrac{\Delta G^\circ}{2.303\, RT}$	$\mathscr{E}^\circ = -\dfrac{\Delta G^\circ}{n\mathscr{F}}$	$\Delta Gh^\circ = \Delta G^\circ$
Nature of transfer potential	$\propto \Delta G^\circ$ per mole H^+ transferred	$\propto \Delta G^\circ$ per mole e transferred	$\propto \Delta G^\circ$ per mole PO_4 transferred

transfer potentials are disguised, or derivative, forms of the free energy. It may be illuminating, therefore, to compare certain common features of these three types of transfer potential (Table 6.1).

In all cases the potential refers to some chemical transformation, usually written in a very concise form. For example, we generally write the reaction

$$AH \rightleftharpoons A^- + H^+ \tag{6.1}$$

in discussions of *acidity*. However, it is necessary to realize that this equation is only a short-hand notation. If taken literally as meaning the removal of a proton from the acid AH, then reaction (6.1), for all practical purposes would never occur; for example, to remove a proton from NH_4^+ to give NH_3 requires an *input* of energy of some 300,000 cal/mole (in the gas phase). The actual meaning of Eq. (6.1) to chemists is not its literal one, however. Rather it is recognized that the proton is actually *transferred* from AH to some acceptor species, for example, water:

$$AH + H_2O \rightleftharpoons A^- + H_3O^+. \tag{6.2}$$

For this transfer reaction the energy change is usually of the order of a few kilocalories (positive or negative), and the transfer does occur to a greater or lesser extent depending on the magnitude of ΔG.

Similarly for electron transfer potentials, *redox potentials*, we usually write the concise equation

$$A \rightleftharpoons A^+ + e \tag{6.3}$$

or, in the now almost universal convention, in the reverse direction. Again, as written, Eq. (6.3) implies that it is describing the ionization process, and this would rarely happen in circumstances of biochemical interest, for the input of energy required to remove an electron from, for example, an Fe^{2+} atom to give Fe^{3+} is of the order of 10^5 cal. In discussions of redox potentials, however, the student at an early stage is made aware that implicit in Eq. (6.3) is the presence of an acceptor, and that the actual reaction is described better by the equation

$$A + H^+ \rightleftharpoons A^+ + \tfrac{1}{2}H_2. \tag{6.4}$$

For this electron transfer process, the energy change is again of the order of a few kilocalories (positive or negative), and the reaction does occur to a greater or lesser extent depending on the magnitude of ΔG.

Likewise, in connection with biochemical *bond energies* we must make a corresponding distinction between the concise representation and the more detailed description of the process implied. The term *bond energy* has a very definite meaning in the field of energetics, quite different from that implied by the biochemical term *high-energy bond*. The outstanding example of a compound with a high phosphate-transfer potential is adenosine triphosphate, in which the "high-energy bond" is symbolized by $\sim$:

$$\text{AD–Rib–P} \sim \text{P} \sim \text{P}.$$

The implication of the name *high-energy bond* is that there is a concentration of energy between $P \sim P$ that tends to make the terminal phosphate fly off whenever possible; however, the $P \sim P$ bond will not spontaneously spring open. Actually, we would have to put energy *into* ATP (approximately 10^5 cal/mole) to remove the terminal phosphate group or, more explicitly, to break the distal P–O–P bonding:

$$\text{Ad}-\text{Rib}-\text{O}-\underset{\underset{O^-}{|}}{\overset{\overset{O}{\|}}{P}}-\text{O}-\underset{\underset{O^-}{|}}{\overset{\overset{O}{\|}}{P}}-\text{O}-\underset{\underset{O^-}{|}}{\overset{\overset{O}{\|}}{P}}-\text{O}^-.$$

This energy which must be forced into the molecule to break a bond between two atoms is what should be called the bond energy* (Fig. 6.1).

On the other hand, the biochemist is not really interested in the change in internal energy necessary to break the P–O bond, but rather in the change in chemical potential (or free energy) when a compound such as

---

* Strictly speaking, this reaction should be carried out in the gaseous state if we wish to call ΔE the bond energy.

A. Comparison of Transfer Potentials

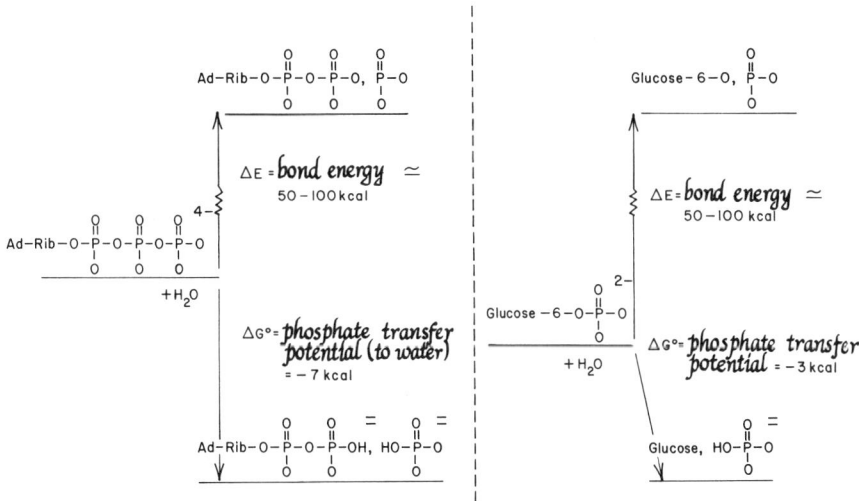

Fig. 6.1. Distinction between bond energy and group transfer potential.

ATP transfers one of its substituent groups to another molecule. Here ΔG for such a transfer reaction obviously ought to depend on the nature of the acceptor molecule as well as on the character of the group donor. Since we are primarily interested in *comparing* group-transfer potentials of a variety of donor molecules, it becomes convenient to select some standard acceptor molecule. This is usually H_2O. Thus, as is shown in Fig. 6.1, the change in chemical potential when the terminal phosphate group is transferred from ATP^{4-} to H_2O, with the concomitant formation of ADP^{2-} and HPO_4^{2-}, is -7 kcal/mole. On the other hand, when glucose 6-phosphate^{2-} transfers its phosphate group to water, $\Delta G°$ is only -3 kcal/mole. Since a negative ΔG signifies that the reaction under consideration can occur spontaneously, it is clear that ATP has a substantially greater phosphate-transfer potential than does glucose 6-phosphate.

Thus, the group-transfer potential may be defined as the change in chemical potential, $\Delta G°$ when one mole of a substituent group of a donor molecule is transferred to a standard acceptor (usually H_2O) under standard concentration conditions. The proper unit for designation of group-transfer potentials is calories (or kilocalories) per mole of transferred group.

Group-transfer potentials for protons and electrons also fundamentally must be related to the free energy changes for the corresponding transfer

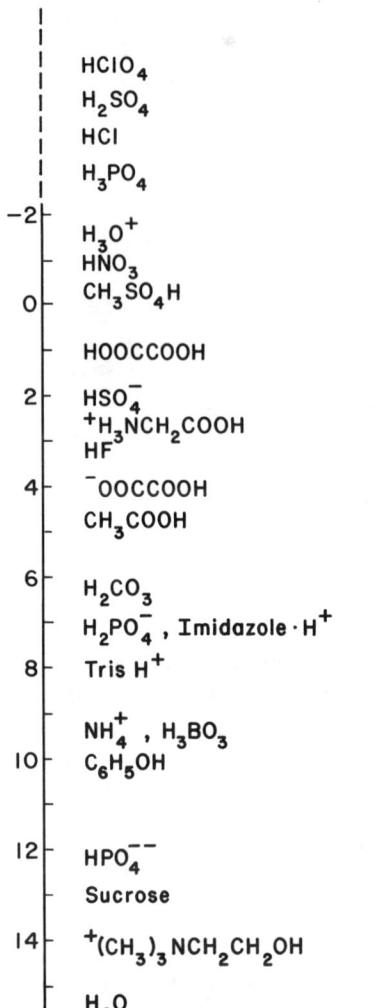

Fig. 6.2. pK_a values for a range of acids from the very strongest (negative pK_a) to very weakest (large positive pK_a).

processes. However, historically, other quantities that served as quantitative measures of the extent of transfer came into general use before their relation to free energy was widely appreciated. Nevertheless, these quantities, pK_a for acidity and $\mathscr{E}°$ for redox potentials, must be proportional to $\Delta G°$. Thus, it follows from Eq. (4.1) that

$$pK_a = \Delta G°/2.303RT, \tag{6.5}$$

where pK_a refers to the reaction represented concisely by Eq. (6.1). Since

A. Comparison of Transfer Potentials

a high pK_a implies a large (positive) $\Delta G°$, the dissociation of a proton would not be favored, and the acid would be weak. On the contrary, a lower pK_a implies a stronger acid. The pK_a transfer potential is thus proportional to $\Delta G°$ for the transfer of a mole of protons from the acid to a reference acceptor molecule [Eq. (6.2)] (Fig. 6.2).

Similarly $\mathscr{E}°$ is proportional to $\Delta G°$. Thus, it follows from Eq. (4.7) that

$$\mathscr{E}° = -\Delta G°/n\mathscr{F}. \tag{6.6}$$

By general convention among biochemists and most physical chemists, $\mathscr{E}°$ is agreed upon to refer to the reverse of Eq. (6.3); in other words, tables of $\mathscr{E}°$ values apply to the process.

$$e + A^+ \rightleftharpoons A. \tag{6.7}$$

Thus, if $\mathscr{E}°$ is a positive number, $\Delta G°$ is negative and the uptake of an electron by A^+ occurs spontaneously. Furthermore, the more positive is $\mathscr{E}°$, the stronger oxidizing agent is A^+ (Fig. 6.3). Thus, the $\mathscr{E}°$ transfer

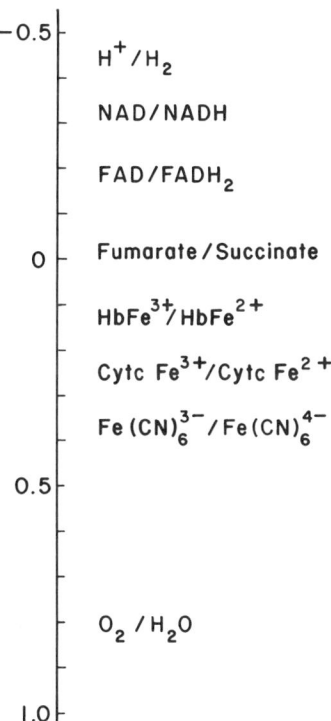

Fig. 6.3. Reduction potentials $\mathscr{E}°$ (in volts) for $e + A^+ \rightarrow A$, for a series of oxidation–reduction couples. The better oxidizing agents are at the lower end of the chart.

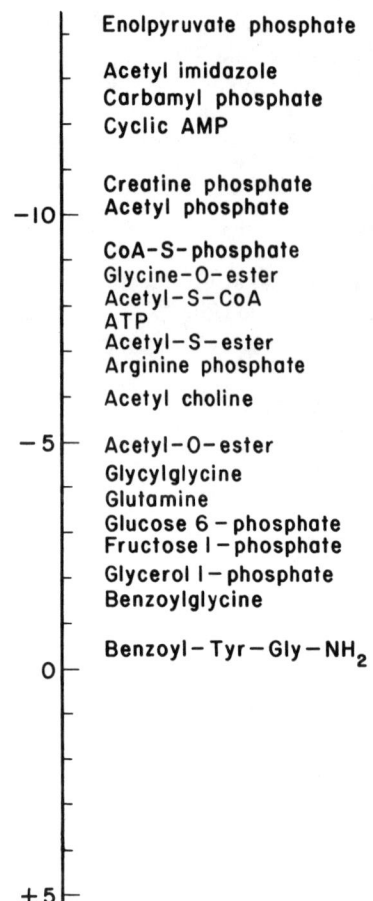

Fig. 6.4. Group transfer potentials (in kilocalories per mole) at pH 7, 25°C.

potential is proportional to $\Delta G°$ for the transfer of a mole of electrons from an acceptor to the substance of interest:

$$\tfrac{1}{2} H_2 + A^+ \rightleftharpoons H^+ + A. \tag{6.8}$$

For group-transfer potentials, $\Delta G°$ itself is the measure of the tendency to transfer a mole of the group (Figure 6.4).

B. COUPLED REACTIONS

The primary biological usefulness of $\mathscr{E}°$ is for decisions as to whether certain oxidation–reduction reactions are feasible within the principles of energetics. Likewise, the main use of group-transfer potentials is to decide

B. Coupled Reactions

whether particular transfer reactions are compatible with thermodynamic requirements. For example, in the conversion of arginine to arginine phosphate,

$$\underset{\substack{\|\\H_2N-C-NH-(CH_2)_3-CH-COO^-}}{\overset{NH_2^+NH_3^+}{}} + HPO_4^{2-} \rightleftharpoons$$

$$\underset{\substack{\|\\{}^{2-}O_3P-NH-C-NH-(CH_2)_3-CH-COO^-}}{\overset{NH_2^+NH_3^+}{}} + H_2O,$$

$$\Delta G° = +7 \text{ kcal/mole} \tag{6.9}$$

the chemical potential must be increased by 7 kcal/mole. Clearly, then, this reaction cannot occur by itself [at standard concentrations (1 M)]. If, however, another reaction can take place concurrently with a $\Delta G°$ of -7 kcal/mole or more negative, then the net reaction would be thermodynamically feasible. For example, we can combine the following two reactions

$$ATP + H_2O \rightleftharpoons ADP + HPO_4^{2-},$$
$$\Delta G° = -7 \text{ kcal/mole}, \tag{6.10}$$

$$\text{Arginine} + HPO_4^{2-} \rightleftharpoons \text{Arginine-P} + H_2O, \tag{6.11}$$
$$\Delta G° = +7 \text{ kcal/mole}$$

to obtain by addition

$$ATP + \text{Arginine} \rightleftharpoons ADP + \text{Arginine-P},$$
$$\Delta G° = 0. \tag{6.12}$$

The resultant reaction will occur to an appreciable extent since, if the standard ΔG is zero, the equilibrium concentrations of ADP and arginine phosphate will be 1 M in the presence of 1 M ATP and 1 M arginine.

On the other hand, addition of the following two reactions

$$\text{Glucose 6-P} + H_2O \rightleftharpoons HPO_4^{2-},$$
$$\Delta G° = -3 \text{ kcal/mole}, \tag{6.13}$$

$$\text{Arginine} + HPO_4^{2-} \rightleftharpoons \text{Arginine-P} + H_2O,$$
$$\Delta G° = +7 \text{ kcal/mole}, \tag{6.11}$$

leads to

$$\text{Glucose 6-P} + \text{Arginine} \rightleftharpoons \text{Glucose} + \text{Arginine-P},$$
$$\Delta G° = +4 \text{ kcal/mole}, \tag{6.14}$$

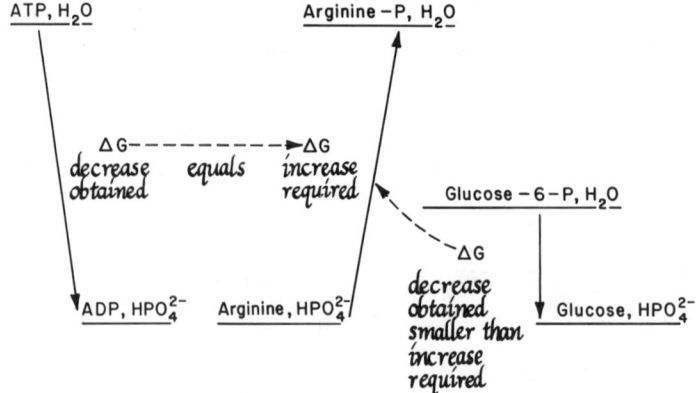

Fig. 6.5. Coupling of potentials or "energy coupling."

for which the net change in chemical potential is a substantial positive number. Thus, reaction (6.14) cannot occur (under standard conditions).

Consequently we might say, speaking rather loosely, that when reaction (6.10) is "coupled" with (6.11), the phosphorylation of arginine is made to proceed, but when reaction (6.13) is "coupled" with (6.11), the phosphorylation of arginine nonetheless cannot be produced. Alternatively (Fig. 6.5) we might recognize that the phosphate-transfer potential of ATP is high and, hence, adequate to place a phosphate group on arginine despite the counter-potential of almost the same magnitude of arginine phosphate.* On the other hand, the phosphate-transfer potential of glucose 6-phosphate is only -3 kcal/mole, which is inadequate to place a phosphate group on arginine since the back-potential of arginine phosphate is greater (-7 kcal/mole).

Coupling of potentials is not limited to reactions involving transfer of the same group in both cases. For example, the free energy liberated when a pyrophosphate group is transferred from ATP to water, with the concomitant formation of adenosine monophosphate (AMP) and pyrophosphate (PP), may be used to supply the free energy required to form an acetylthiol ester of high transfer potential. The individual reactions.[†]

$$ATP + H_2O \rightleftharpoons AMP + PP,$$
$$\Delta G° = -7 \text{ kcal mole}^{-1},$$
(6.15)

---

* The phosphate-transfer potential of arginine phosphate is -7 kcal/mole, as can be seen by changing reaction (6.11) to read from right to left.
† HS—CoA represents Coenzyme A.

C. Experimental Determination of Group-Transfer Potentials

$$CH_3COOH + HS-CoA \rightleftharpoons CH_3COS\text{-}CoA + H_2O,$$

$$\Delta G° = +7 \text{ kcal/mole},\tag{6.16}$$

if coupled, lead to

$$ATP + CH_3COOH + HS-CoA \rightleftharpoons AMP + PP + CH_3COS-CoA;$$

$$\Delta G° \simeq 0.\tag{6.17}$$

The value of $\Delta G°$ indicates that equilibrium can be attained when substantial quantities (1 M concentrations) of the products, as well as of the reactants, are present. Thus, appreciable quantities of acetyl-S-CoA are formed at equilibrium. Since acetyl-S-CoA has a high tendency to transfer its acetyl group to other acceptors, we have formed a compound of high acetyl-transfer potential from one of high phosphate-transfer potential (ATP). Similarly, we can produce a compound with high methyl-group transfer potential using the free energy obtainable from the hydrolytic splitting of the phosphate groups of high transfer potential in ATP.

Considerations of relative group-transfer potentials (Fig. 6.4) tell us whether a particular combination of reactions is thermodynamically feasible. An affirmative answer does not mean, however, that the molecular coupling, or the actual molecular mechanism of the transformation, is the same as the reactions whose potentials have been coupled. The resultant ΔG of a reaction depends only on the initial and final reactants and on their states but not on the actual molecular mechanism through which the reaction occurs. Thus, for reaction (6.17) $\Delta G°$ is zero whether the reaction occurs through steps (6.15) and (6.16) or (as is known to be the case) through a more complicated sequence of steps. Classical energetics, since it is independent of the assumptions of atomic theory, can give no information on mechanisms at the molecular level.

C. EXPERIMENTAL DETERMINATION OF GROUP-TRANSFER POTENTIALS

We shall examine in some detail the experimental data that must be accumulated to evaluate the power of a particular high-energy bond. In the course of these computations we shall also specify the nature of the reactions more precisely.

1. Phosphate Bond Energy

For precise calculations we must recognize at the outset that at physiological pH, ionizing molecules frequently exist in more than one (charged) form. Thus in ATP,

$$\text{Adenine—Ribose—O—}\underset{\underset{\text{O}^-}{|}}{\overset{\overset{\text{O}}{\|}}{\text{P}}}\text{—O—}\underset{\underset{\text{O}^-}{|}}{\overset{\overset{\text{O}}{\|}}{\text{P}}}\text{—O—}\underset{\underset{\text{O}^-}{|}}{\overset{\overset{\text{O}}{\|}}{\text{P}}}\text{—OH},$$

for example, three of the phosphate hydroxyl groups have pK_a values near 1–2 and are completely ionized at pH's around 7, but the fourth hydroxyl has a pK_a value of 6.50 and, hence, is only partially dissociated at most pH's of biochemical interest. Before writing an equation for the hydrolysis of ATP, we should specify the charged forms of each reactant and product. In concise notation these may be represented by the following equilibria:

$$ATP^{3-} \rightleftharpoons ATP^{4-} + H^+, \quad K_\alpha = 10^{-6.50}, \tag{6.18}$$

$$ADP^{2-} \rightleftharpoons ADP^{3-} + H^+, \quad K_\beta = 10^{-6.27}, \tag{6.19}$$

$$H_2PO_4^- \rightleftharpoons HPO_4^{2-} + H^+, \quad K_\gamma = 10^{-6.73}. \tag{6.20}$$

Thus, the hydrolysis of ATP at pH 7, for example, really should be represented by

$$\sum ATP + H_2O \rightleftharpoons \sum ADP + \sum PO_4, \tag{6.21}$$

where

$$[\sum ATP] = [ATP^{3-}] + [ATP^{4-}], \tag{6.22}$$

$$[\sum ADP] = [ADP^{2-}] + [ADP^{3-}], \tag{6.23}$$

$$[\sum PO_4] = [H_2PO_4^-] + [HPO_4^{2-}]. \tag{6.24}$$

The standard free energy change $\Delta G°$ for reaction (6.21) refers, therefore, to ΔG for *total* concentration of ATP (Σ ATP), at unit activity, and for the *total* concentrations of ADP and PO$_4$, respectively, at unit activity.

To actually evaluate this $\Delta G°$ we turn to experimental measurements of equilibrium constants. The equilibrium constant for Eq. (6.21) lies so far to the right, that is, the hydrolysis of ATP is so nearly complete at equilibrium, that K_Σ,

$$K_\Sigma = [\sum ADP][\sum PO_4]/[\sum ATP], \tag{6.25}$$

cannot be evaluated directly. However, we can couple this reaction with another, the synthesis of glutamine (Gln) from glutamic acid (Glu$^-$) and ammonium ion,

$$Glu^- + NH_4^+ \rightleftharpoons Gln + H_2O, \tag{6.26}$$

C. Experimental Determination of Group-Transfer Potentials

for which the equilibrium constant K_G is known:

$$K_G = [\text{Gln}]/[\text{Glu}^-][\text{NH}_4^+] = 0.00315 \quad \text{(at pH 7, 310° K)}. \quad (6.27)$$

The coupled reaction

$$\text{Glu}^- + \text{NH}_4^+ + \sum \text{ATP} \rightleftharpoons \text{Gln} + \sum \text{ADP} + \sum \text{PO}_4 \quad (6.28)$$

will attain equilibrium in the presence of an enzyme, glutamine synthetase, and the equilibrium constant K' has been measured:

$$K' = \frac{[\text{Gln}][\sum \text{ADP}][\sum \text{PO}_4]}{[\text{Glu}^-][\text{NH}_4^+][\sum \text{ATP}]} = 1200 \quad \text{(at pH 7, 310° K)}. \quad (6.29)$$

Inspection of Eqs. (6.25), (6.27), and (6.29) shows that

$$K' = K_\Sigma K_G. \quad (6.30)$$

Therefore,

$$K_\Sigma = 3.8 \times 10^5, \quad (6.31)$$

and

$$\Delta G^{\circ\prime} = -RT \ln K_\Sigma = -7900 \quad \text{cal/mole} \quad (6.32)$$

at 310 K. Thus, the phosphate bond energy, or transfer potential, in ATP at pH 7 (and 37°C) is -7.9 kcal/mole.

2. Variation of Transfer Potential with Standard Species Chosen

In the preceding section we have considered the bond energy of ATP under "practical" conditions, that is, in a solution in which various ionic species exist. For some purposes we might wish to know the intrinsic transfer potential of a single charged species going to products in only one ionic state each. For example, we might consider the reaction

$$\text{ATP}^{4-} + \text{H}_2\text{O} \rightleftharpoons \text{ADP}^{2-} + \text{HPO}_4^{2-}. \quad (6.33)$$

Here ΔG° for this reaction refers to ΔG for ATP^{4-} at a concentration corresponding to unit activity regardless of what the concentration of ATP^{3-} or $\sum \text{ATP}$ might be. Similarly, it refers to concentrations of unit activity for ADP^{2-} and HPO_4^{2-}. Thus, ΔG° for Eq. (6.33) need not be the same as $\Delta G^{\circ\prime}$ for Eq. (6.21). Nevertheless, these two standard ΔG values are related to each other because the corresponding equilibrium constants are interdependent.

Thus, for Eq. (6.33), we may write an equilibrium constant

$$K = [\text{ADP}^{2-}][\text{HPO}_4^{2-}]/[\text{ATP}^{4-}]. \quad (6.34)$$

However, each concentration factor in Eq. (6.34) may be related to a corresponding concentration sum [Σ]. For example, from Eq. (6.22) we have

$$[\text{ATP}^{4-}] = [\textstyle\sum \text{ATP}] - [\text{ATP}^{3-}]. \tag{6.35}$$

If we also make use of the information in Eq. (6.18), we obtain

$$[\text{ATP}^{3-}] = [\text{H}^+][\text{ATP}^{4-}]/K_\alpha, \tag{6.36}$$

which can be inserted into Eq. (6.35) and the resultant rearranged to give

$$[\text{ATP}^{4-}] = [\textstyle\sum \text{ATP}]\frac{K_\alpha}{[\text{H}^+] + K_\alpha}. \tag{6.37}$$

Similar manipulations with Eqs. (6.19) and (6.23) lead to

$$[\text{ADP}^{2-}] = [\textstyle\sum \text{ADP}]\frac{[\text{H}^+]}{[\text{H}^+] + K_\beta}, \tag{6.38}$$

and Eqs. (6.20) and (6.24) give

$$[\text{HPO}_4^{2-}] = [\textstyle\sum \text{PO}_4]\frac{K_\gamma}{[\text{H}^+] + K_\gamma}. \tag{6.39}$$

Equations (6.37)–(6.39) may then be substituted into Eq. (6.34), and we arrive at the relationship

$$K = \frac{[\sum \text{ADP}][\sum \text{PO}_4]}{[\sum \text{ATP}]}\left[\left(\frac{[\text{H}^+]}{[\text{H}^+]+K_\beta}\frac{K_\gamma}{[\text{H}^+]+K_\gamma}\right)\bigg/\frac{K_\alpha}{[\text{H}^+]+K_\alpha}\right] \tag{6.40}$$

or

$$K = K_\Sigma\left[\left(\frac{[\text{H}^+]}{[\text{H}^+]+K_\beta}\frac{K_\gamma}{[\text{H}^+]+K_\gamma}\right)\bigg/\frac{K_\alpha}{[\text{H}^+]+K_\alpha}\right]. \tag{6.41}$$

We have experimentally evaluated K_Σ at $[\text{H}^+] = 10^{-7}$, and we know the values of the ionization constants K_α, K_β, and K_γ. Clearly then, K can be computed from the appropriate arithmetic manipulations required by Eq. (6.41). The result obtained is

$$K = 5.1 \times 10^3, \tag{6.42}$$

and hence $\Delta G°$ for reaction (6.33) is

$$\Delta G° = -6700 \quad \text{cal/mole}. \tag{6.43}$$

Thus, the transfer potential of phosphate form ATP^{4-} to form ADP^{2-} and HPO_4^{2-} is significantly different than the phosphate bond energy in the "practical" reaction at pH 7.

C. Experimental Determination of Group-Transfer Potentials

Equation (6.41) may also be used in a complementary fashion, to evaluate $\Delta G^{\circ\prime}$ (from K_Σ) at pH's other than 7. That this is possible may be understood from the following considerations. The ΔG° of -6.7 kcal/mole of Eq. (6.43) is the free energy change for the chemical reaction of Eq. (6.33) in which

$$[\text{ATP}^{4-}] = 1; \quad [\text{ADP}^{2-}] = 1; \quad [\text{HPO}_4^{2-}] = 1, \qquad (6.44)$$

i.e., in which each substance is at unit concentration (assuming activity can be replaced by concentration alone). So long as the standard concentration conditions of each species of Eq. (6.44) are retained, ΔG° is -6.7 kcal/mole, no matter what the pH. At different pH's, the relative amounts of $[\text{ATP}^{4-}]$ and $[\text{ATP}^{3-}]$ will vary, but if we maintain $[\text{ATP}^{4-}]$ at unit concentration it will hydrolyze (to $[\text{ADP}^{2-}]$ and $[\text{HPO}_4^{2-}]$ each at unit concentration) with the same ΔG° of -6.7 kcal/mole. Therefore, K of Eq. (6.42), being directly proportional to ΔG° of Eq. (6.43), will be a *constant* at all pH's.

Keeping this conclusion in mind and looking again at Eq. (6.41), we see that the left-hand side is a fixed number at all pH's. On the right-hand side, the bracketed factor contains $[\text{H}^+]$ and, therefore, has a different value at different pH's. Consequently K_Σ must vary in a compensating manner with change in pH. Therefore $\Delta G^{\circ\prime}$ [Eq. (6.32)] depends on pH, or in other words, the free energy change for the hydrolysis reaction of Eq. (6.21) varies with pH. That this should be so is not unreasonable since the proportion of ATP^{4-} and ATP^{3-} varies with pH even if $[\Sigma \text{ATP}]$ is kept fixed at 1, and ATP^{4-} has a different ΔG° of hydrolysis than ATP^{3-}. The actual numerical value of $\Delta G^{\circ\prime}$ for reaction (6.21) at different pH's can then be computed from the K_Σ at the corresponding pH, K_Σ being calculated readily from Eq. (6.41).

3. Thioester Bond Energy

Another example of an experimental evaluation of a group-transfer potential using a different approach from that for the phosphate bond energy may be illustrated with $\text{CH}_3\text{CO-SR}$. The heats of hydrolysis of a number of thioesters with different R groups have been measured and are all nearly equal

$$\text{CH}_3\text{CO-SR (l)} + \text{H}_2\text{O (l)} \rightleftharpoons \text{CH}_3\text{COOH (l)} + \text{HSR (l)};$$
$$\Delta H_1^\circ \simeq -1 \text{ kcal/mole}. \qquad (6.45)$$

Similarly for the corresponding oxygen esters

$$\text{CH}_3\text{CO-OR (l)} + \text{H}_2\text{O (l)} \rightleftharpoons \text{CH}_3\text{COOH (l)} + \text{HOR (l)};$$
$$\Delta H_2^\circ \simeq +1 \text{ kcal/mole}. \qquad (6.46)$$

For the first of this pair of reactions we may write

$$\Delta G_1^\circ = \Delta H_1^\circ - T\Delta S_1^\circ, \tag{6.47}$$

and for the second

$$\Delta G_2^\circ = \Delta H_2^\circ - T\Delta S_2^\circ, \tag{6.48}$$

so that for Eq. (6.45) minus Eq. (6.46),

$$CH_3CO-SR + ROH \rightleftharpoons CH_3CO-OR + RSH, \tag{6.49}$$

we obtain

$$\Delta(\Delta G^\circ) = \Delta G_1^\circ - \Delta G_2^\circ = (\Delta H_1^\circ - \Delta H_2^\circ) - T(\Delta S_1^\circ - \Delta S_2^\circ). \tag{6.50}$$

Since the species on the left-hand side of Eq. (6.49) are very similar in molecular arrangement and configuration to the corresponding species on the right-hand side, it is reasonable to assume (see Chapter 8) that there is no significant overall entropy change in this reaction, that is,

$$\Delta S_1^\circ - \Delta S_2^\circ = 0. \tag{6.51}$$

Therefore,

$$\Delta G_1^\circ = \Delta G_2^\circ + (\Delta H_1^\circ - \Delta H_2^\circ). \tag{6.52}$$

We know ΔH_1° and ΔH_2°. For the ester hydrolysis, ΔG_2° has been experimentally evaluated under a variety of conditions. At pH 7 it is -5.1 kcal/mole. Hence,

$$\Delta G_1^\circ = -5.1 + (-1 - 1) = -7.1 \quad \text{kcal/mole}. \tag{6.53}$$

Thus, the thioester group has a relatively high transfer potential.

By methods such as these, group-transfer potentials have been evaluated for a variety of substances. Some "best" values have been assembled in Fig. 6.4. This table may be used in a manner similar to one of redox potentials (Fig. 6.3) or pK_a values (Fig. 6.2). Substances high in the list can spontaneously transfer a group to acceptors lower in the list under standard concentration conditions.

EXERCISES

1. The equilibrium constant for the isomerization reaction

is 56 at 39°C and pH 9.22 [W. P. Jencks, S. Cordes, and J. Carriuolo, *J. Biol. Chem.* **235**, 3608 (1960)]. Using $\Delta G°$ of hydrolysis of ethyl acetate as a measure of the "bond energy" of oxygen esters, find the transfer potential or "bond energy" of the acetylthioester group.

2. The equilibrium in the hydrolysis of glucose 6-phosphate

$$\text{Glucose 6-P} + H_2O \rightleftharpoons \text{glucose} + P$$

has been studied by Meyerhof and Green [*J. Biol. Chem.* **178**, 655 (1948)]. If the equilibrium constant is expressed in terms of total concentration of each species Σ,

$$K_\Sigma = \frac{[\text{glucose}][\Sigma\, P]}{[\Sigma\, \text{G-6-P}][H_2O]},$$

a value of 122 is found at pH 8.5 and 38°C. What is $\Delta G°$ of the following reaction at pH 8.5 and 38°C?

$$\text{Glc}-O-\overset{\overset{O}{\|}}{\underset{\underset{O^-}{|}}{P}}-O^- + H_2O \rightleftharpoons \text{Glc} + HO-\overset{\overset{O}{\|}}{\underset{\underset{O^-}{|}}{P}}-O^-$$

Necessary ionization constants are listed below.

$$GlcPO_4H^- \rightleftharpoons GlcPO_4^{2-} + H^+, \quad K_\alpha = 10^{-6.03},$$

$$H_2PO_4^- \rightleftharpoons HPO_4^{2-} + H^+, \quad K_\beta = 10^{-6.73}.$$

Answer: $\Delta G° = -2980$ cal/mole.

3. The enzyme phosphoglucomutase catalyzes the reaction

$$(\alpha \text{ and } \beta)\text{-D-glucose-6}-O-\overset{\overset{O}{\|}}{\underset{\underset{O^-}{|}}{P}}-O^- \rightleftharpoons \alpha\text{-D-glucose-1}-O-\overset{\overset{O}{\|}}{\underset{\underset{O^-}{|}}{P}}-O^- \quad (1)$$

As indicated, the 6-phosphate is a mixture of the α and β forms at the 1-position, but the 1-phosphate is pure α. At pH 7 and 25°C, the equilibrium constant K for reaction (1) as written is 0.059 [M. R. Atkinson, E. Johnson, and R. K. Morton, *Biochem. J.* **79**, 12 (1961)]. Let us assume that the equilibrium proportions of α and β forms of the glucose 6-phosphate are the same as in ordinary glucose, for which at equilibrium α is 50% of the total.

Find $\Delta G°$ at pH 7 and 25°C for the reaction:

$$\alpha\text{-D-glucose-6-phosphate} \rightleftharpoons \alpha\text{-D-glucose-1-phosphate}. \quad (2)$$

Answer: $\Delta G° = +1140$ cal/mole.

4. The acetyl group transfer potential at 25°C and pH 7 in acetyl-Coenzyme A, $CH_3CO—S—CH_2CH_2NHR$, can be computed from the following information provided by Jencks and Gilchrist [W. P. Jencks and M. Gilchrist, *J. Am. Chem. Soc.* **86**, 4657 (1964)]:

$$CH_3\overset{O}{\overset{\|}{C}}—O—CH_2CH_2—N(CH_3)_3^+ + H_2O \rightleftharpoons CH_3\overset{O}{\overset{\|}{C}}—OH + HOCH_2CH_2—N(CH_3)_3^+,$$

$$\Delta G° = 2940 \quad \text{cal/mole},$$

$$CH_3\overset{O}{\overset{\|}{C}}—O—CH_2CH_2—N(CH_3)_3^+ + HS—CH_2CH_2NHR$$

$$\rightleftharpoons CH_3\overset{O}{\overset{\|}{C}}—S—CH_2CH_2NHR + HOCH_2CH_2—N(CH_3)_3^+,$$

$$K = 0.076$$

K_a for CH_3COOH is 1.75×10^{-5}. What is the "bond energy" of the acetyl—S bond in acetyl—CoA?

5. The free energy of hydrolysis of acetic anhydride (Ac_2O) to uncharged acetic acid (AcOH) is $-15,700$ cal/mole at 25°C. [W. P. Jencks, F. Barley, R. Barnett, and M. Gilchrist, *J. Am. Chem. Soc.* **88**, 4464 (1966).] The equilibrium constant for the formation of acetic anhydride and imidazole (Im) from acetylimidazole (AcIm) and AcOH is 3.2×10^{-5} in aqueous solution at 25°C. Is the Ac–Im bond a high-energy one? Give a quantitative basis for your answer.

6. Heats of reaction for the transfer of methyl groups from a variety of compounds to a common acceptor, homocysteine, have been determined by Mudd et al. [S. H. Mudd, W. A. Klee, and P. D. Ross *Biochemistry* **5**, 1653 (1966)] and are tabulated below.

Methyl donor	ΔH(kcal/mole)
$(CH_3)_2S^+CH_3$	-6.8
$(CH_3)_2S^+CH_2COO^-$	-10.6
$(CH_3)_2S^+CH_2CH_2COO^-$	-8.7
$(CH_3)_2S^+CH_2CHCHCOO^-$ $\|$ NH_3^+	-6.7
$CH_3S^+CH_2CH_2CHCOO^-$ $\|\quad\quad\|$ $\quad\quad NH_3^+$ 5′-deoxyadenosine	-13.2

Exercises

The reasonable assumption can be made that $\Delta S°$ for the transfer of CH_3 from one compound to another in this series is near zero. Thus, the values of ΔH listed may also serve as good approximations of the methyl group-transfer potentials. It would be desirable, however, to check these assumptions, for example, by making an experimental measurement of an equilibrium constant for one of the methyl transfer reactions.

(a) Which pair of substances in the table would seem most suitable for such an experimental measurement?

(b) What value would you predict for the equilibrium constant if you use the data in the preceding table?

7. If the phosphate bond energy of ATP in a cell internally at pH 7 (and 37°C) is used to expel protons to the external milieu, what is the lowest pH that can be created externally?

Answer: pH 1.5

7

Some Laws of Physicochemical Behavior

Part of the aesthetic attractiveness of the theory of energetics lies in its ability to predict a wide variety of scientific laws without any assumptions beyond the two fundamental laws of thermodynamics described in Chapters 1 and 2. These new principles can be deduced logically from the fundamental laws, most conveniently through the free energy function. We shall not consider the rigorous, logical steps in these derivations, but it is worthwhile to obtain a general idea of the approach used. In this way, we may develop an appreciation of the power of thermodynamic methods in the analysis of physicochemical behavior.

A. ELECTROCHEMICAL RELATIONSHIPS

The primary problem in the thermodynamic analysis of an electrochemical cell is to establish the chemical or physical change that occurs as a result of the flow of current. These details are not always easy to obtain,

A. Electrochemical Relationships

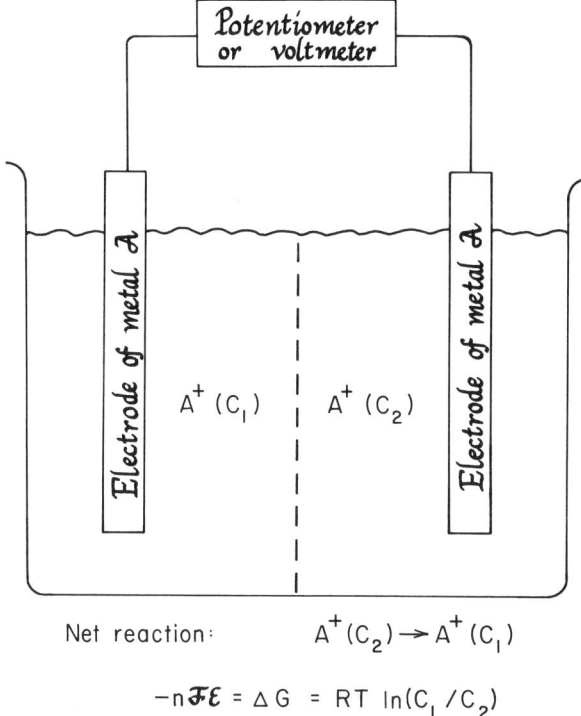

Fig. 7.1. Electrochemical cell: concentration type.

particularly when liquids of different composition are in contact with each other, as is so often the case in biological systems.

Let us consider a simple cell, such as is shown schematically in Fig. 7.1. An electrode of metal A is dipped into one solution of a salt of A, containing A^+ ions at a concentration C_1. An identical electrode is inserted into the right-hand chamber of the cell containing another solution of the same salt of A but at a concentration C_2. The two liquids are kept separated, but in electrical contact, by means of a suitable porous membrane, or, still better, by an inverted U-tube bridge containing concentrated KCl.

We must now determine the chemical changes that would occur if one mole of electrons of current were to flow through the circuit. The electron current can be assumed to flow from left to right or right to left; for our analysis it does not matter. We shall adopt the common convention, therefore, of considering the electrons to flow in the outer circuit, through the potentiometer, from left to right. For this to occur, electrons must be produced at the lefthand electrode; they can be liberated from the metal in

Net reaction: $A^+(C_2) \rightarrow A^+(C_1)$

$-n\mathcal{F}\mathcal{E} = \Delta G = RT \ln(C_1/C_2)$

the reaction

$$A \rightleftharpoons A^+(C_1) + e. \tag{7.1}$$

These electrons could then flow through the potentiometer toward the right-hand electrode; here they would have to be absorbed, or they would pile up on the metal and produce an enormous electrostatic repulsion. The absorption reaction is

$$e + A^+(C_2) \rightleftharpoons A. \tag{7.2}$$

To complete the electrical flow, ions in the solutions must carry negative charge from right to left. The concentration changes which these ionic flows produce are often complex, particularly at junctions between different liquids, but in a suitable experimental arrangement the effect of these changes on the net chemical reaction can be made very small. We neglect any liquid-junction effects, therefore, and assume that the net reaction in the electrochemical cell of Fig. 7.1 is the sum of Eqs. (7.1) and (7.2):

$$A^+(C_2) \rightleftharpoons A^+(C_1). \tag{7.3}$$

As for any other chemical or physical transformation, there is a ΔG associated with reaction (7.3). In view of the fact that this net reaction involves merely changing the concentration of A^+ from one value (C_2) to another (C_1), it is evident that we can relate ΔG to the concentrations by means of a relation similar to Eq. (5.3):

$$\Delta G = RT \ln(C_1/C_2). \tag{7.4}$$

In addition, ΔG originally acquired the name "free energy" change because it is also a measure of the maximum useful work which can be obtained from a reaction. More precisely,

$$\Delta G = W_{max} \tag{7.5}$$

if, as in Chapter 1, we assign a positive numerical value to W when work is done on the system. In our present problem this work is electrical in nature, so we may also write

$$W_{max} = -\text{charge} \times \text{potential}.$$

$$= -q \times \mathscr{E}. \tag{7.6}$$

The charge carried by one mole of electrons is 96,500 coulombs, usually abbreviated by the symbol $\mathscr{F}$ (the Faraday), so we may write

$$W_{max} = -\mathscr{F}\mathscr{E}. \tag{7.7}$$

To generalize just a little further, let us also include cases in which the

B. Osmotic Pressure

number of moles of electrons, n, might be more or less than one and write

$$W_{max} = -n\mathscr{F}\mathscr{E}. \tag{7.8}$$

It follows from Eqs. (7.5) and (7.8) that

$$\Delta G = -n\mathscr{F}\mathscr{E}. \tag{7.9}$$

Consequently, since ΔG of Eq. (7.9) refers to the same chemical change [reaction (7.3)] as does ΔG of Eq. (7.4), the two are the same thing, and

$$-n\mathscr{F}\mathscr{E} = RT \ln(C_1/C_2) \tag{7.10}$$

or

$$\begin{aligned}\mathscr{E} &= -(RT/n\mathscr{F})\ln(C_1/C_2) \\ &= (RT/n\mathscr{F})\ln(C_2/C_1).\end{aligned} \tag{7.11}$$

If, for example, A^+ were Na^+ ion, we may write specifically

$$\mathscr{E} = (RT/n\mathscr{F}) \ln [Na^+]_2/[Na^+]_1. \tag{7.12}$$

Thus we arrive at the well-known Nernst equation for electrochemical potentials, in this example, for a concentration cell. The same general procedure may be used for the derivation of equations for the electromotive force of more complicated electrochemical systems. In every case it is necessary first to write the net chemical or physical transformation that occurs in the cell, second to formulate the appropriate expression for ΔG for this transformation, and third to equate this ΔG to the electrical work. Simple algebraic manipulation will lead to the equation for $\mathscr{E}$.

B. OSMOTIC PRESSURE

In analyzing the thermodynamic basis of osmotic phenomena, we also find it convenient to visualize an appropriate idealized system such as is shown in Fig. 7.2. The two arms of a U-tube, separated by a partition consisting of a membrane permeable to the solvent but not to the solute (dashed line in Fig. 7.2), are first filled with pure solvent to equal heights in both arms. Under these conditions, and with the pressure P the same as the atmospheric pressure P_0 and the temperature constant, the system is at equilibrium. Hence,

$$\Delta G = 0 = G_{right} - G_{left}, \tag{7.13}$$

or the chemical potential of solvent on the right is equal to that on the left.

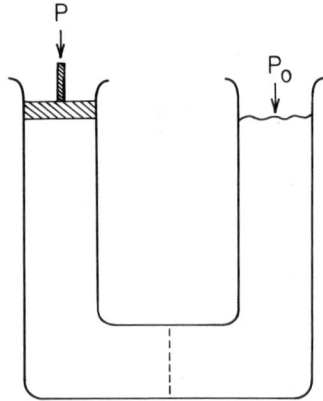

Fig. 7.2. Osmotic pressure. Solute is dissolved in chamber on left, and partition is assumed to be impermeable to solute but permeable to solvent.

Now we add some solute to the left-hand side. The presence of dissolved solute lowers the chemical potential of the solvent by an amount proportional to the number of moles of added solute, n_2:

$$\text{Lowering of } G \text{ of solvent due to solute} = RTn_2. \qquad (7.14)^*$$

In the absence of any other stress, solvent would move from the right side of the tube to the left, since that on the right retains its initial higher chemical potential. Such a flow could be prevented, however, by increasing the chemical potential of the solvent on the left, for example, by making the pressure P greater then the atmospheric pressure P_0:

$$\text{Increase of } G \text{ of solvent due to increased pressure} = V(P - P_0). \qquad (7.15)$$

In this equation, V is the volume of solvent on the left side of the U-tube. If the increase in G due to the extra pressure exerted just balances the decrease due to added solute, then equilibrium will be re-established, no solvent will flow from one side to the other, and we may equate (7.14) and (7.15).

$$V(P - P_0) = RTn_2. \qquad (7.16)$$

This increase in pressure necessary to maintain equilibrium in the presence of added solute is named the osmotic pressure π:

$$\pi = P - P_0. \qquad (7.17)$$

Making use of Eq. (7.17) and rearranging Eq. (7.16) we obtain the familiar form

$$\pi = RTn_2/V = RTC_2, \qquad (7.18)$$

where C_2 is the number of moles of dissolved solute per liter.

*We accept Eqs. (7.14) and (7.15) without proof.

Since π depends on C_2, if we can measure the osmotic pressure, we can calculate C_2 in moles per liter. Thus, osmotic pressure measurements provide a means of obtaining the molecular weight of a solute. In practice this method is particularly useful for macromolecules.

C. MOLECULAR WEIGHT FROM ULTRACENTRIFUGATION

One of the earliest of the physicochemical characteristics that we wish to know about a biological macromolecule is its molecular weight. Ultracentrifugation is perhaps the most versatile method for obtaining such weights.

The ultracentrifuge may be used in several different ways to obtain molecular weights. One of these procedures is known as *sedimentation equilibrium*. A solution is placed in a small cell which in turn is set in a heavy rotor [Fig. 7.3a]. The rotor is then rotated at high speed ($\sim 15,000$ rev/min) so that a strong centrifugal force is produced which

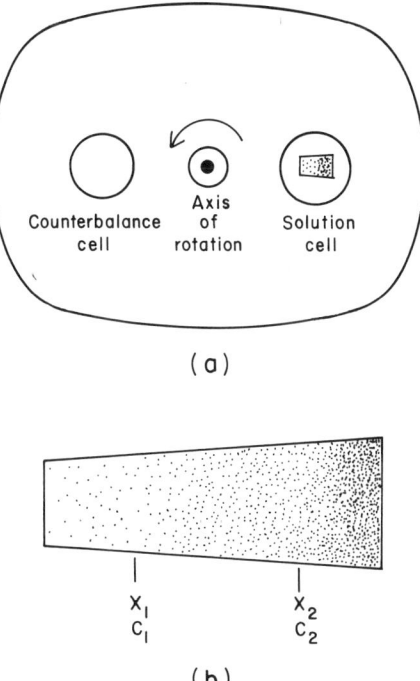

Fig. 7.3. Sedimentation equilibrium: (a) heavy rotor with solution cell shown and (b) distribution of solute in direction of centrifugal field.

tends to move the molecules toward the outer end of the cell [Fig. 7.3b]. As the molecules congregate at the right side of the cell, their concentration increases, but in turn the concentration of those remaining at the left is diminished. As a result, a *diffusional force* is set up, opposite in direction to the centrifugal force, which tends to move molecules back from right to left. Ultimately the centrifugal and diffusional forces balance exactly and equilibrium is established in which a definite concentration gradient, increasing from left to right, exists in the cell indefinitely. At this stage, the concentrations C_1 and C_2 are measured at two points in the cell at distances x_1 and x_2 from the center of the rotor. From this information (usually obtained by optical means) and the following theoretical analysis, we can compute the molecular weight.

Since the concentrations C_1 and C_2 differ, there must be a difference in free energy between solute at x_1 and that at x_2 due to this concentration gradient. We can compute this ΔG readily from Eq. (5.1), in the same way that we have just obtained Eq. (7.4):

$$\Delta G_{\text{conc}} = G_1(\text{at } x_1) - G_2(\text{at } x_2) = RT \ln C_1 - RT \ln C_2$$
$$= RT \ln(C_1/C_2). \tag{7.19}$$

Since C_1 is less than C_2, C_1/C_2 will be less than unity, its logarithm will be negative, and hence ΔG will be negative. It follows, therefore, that molecules ought to move back spontaneously from position x_2 to x_1. We know that they do not do so. Evidently the concentrational ΔG must be exactly balanced by a free energy change between x_1 and x_2 due to the centrifugal field.

This $\Delta G_{\text{centrifugal}}$ is readily obtained from the relationship between free energy and maximum work, as was the case with the electrochemical problem earlier. To move an object of mass M against a centrifugal force (due to an acceleration of $\omega^2 x$) from point x_2 to x_1 requires an amount of work

$$W = \tfrac{1}{2} M \omega^2 (x_1^2 - x_2^2), \tag{7.20}*$$

where ω is the angular velocity of the rotor. If the object is in a solvent, its effective mass will not be M, but will be less because of a buoyancy correction. This correction, as we may recall from Archimedes' principle, is the weight of the medium displaced by the solute. If we know the volume of 1 g of solute, v, then $v\rho$, where ρ is the density of the solution, gives the weight of solvent displaced by 1 g of solute. If $v\rho$ approaches unity, then the effective mass of the solute $(1 - v\rho)$ reaches zero, and the solute will

---

* We accept Eq. (7.20) without proof.

C. Molecular Weight from Ultracentrifugation

float in the solvent no matter how strong the centrifugal field. Thus, it should become apparent that with a buoyancy correction, the centrifugal work done by the object of mass M moving outward from the axis of rotation is

$$W = -\tfrac{1}{2}M(1 - v\rho)\omega^2(x_2^2 - x_1^2). \tag{7.21}$$

Returning then to our thermodynamic requirement that in going from position x_1 to x_2 at equilibrium

$$\Delta G_{\text{conc}} + \Delta G_{\text{cent}} = 0 \tag{7.22}$$

and reminding ourselves of Eq. (7.5), we may write

$$RT \ln(C_2/C_1) = \tfrac{1}{2}M(1 - v\rho)\omega^2(x_2^2 - x_1^2). \tag{7.23}$$

Thus, for an explicit equation for the molecular weight, we obtain

$$M = \frac{2RT \ln(C_2/C_1)}{(1 - v\rho)\omega^2(x_2^2 - x_1^2)}. \tag{7.24}$$

We have thus derived laws of physicochemical behavior for electrochemical cells, osmotic pressure, and sedimentation. In each case our procedure was to consider the balancing free energies involved in a given system. For each type of free energy we obtained a relationship between ΔG and experimentally measurable quantities. Thus, we found equations relating ΔG to concentration differences, electric potentials, pressure differences, and centrifugal fields. Equating these relationships for corresponding balancing ΔG's, we obtained a new equation that must govern the behavior of the system considered.

There are also relationships, which we have not considered, between ΔG and temperature, tension, surface properties, magnetic properties, etc. It should thus be evident that from simply the two basic laws of thermodynamics, through the concept of free energy, we can derive a host of principles that govern the behavior of matter.

8

Energetics From a Molecular-Statistical Viewpoint

In classical energetics we derive a series of relationships, based on the two fundamental laws of thermodynamics, which correlate particular thermodynamic functions (e.g., the chemical potential) with experimentally measurable properties of matter in bulk (e.g., the equilibrium constant). Since the establishment of the laws of thermodynamics about a century ago, our understanding of the properties of matter in bulk from the atomic standpoint has developed immensely. We might reasonably expect, therefore, that it should be possible to express thermodynamic functions in terms of atomic and molecular properties. It is this combination of specific molecular models with certain theorems of statistics governing the behavior of large assemblies of individual molecules that makes up the branch of energetics frequently named *statistical thermodynamics*.

A. FUNDAMENTAL ASSUMPTIONS

1. Molecular Assumptions

One fundamental axiom of statistical energetics is that matter is constituted of atoms. Implicit in this assumption are all the attributes and properties that we associate with atoms and molecules. Some of these details, such as the internal constitution of atoms, can be ignored for our present purposes. Others, particularly the quantum characteristics of molecular systems, must be examined more carefully.

The basic principle of quantum theory is that the energy of an atom or molecule cannot vary continually but may possess only certain discrete, very sharp, or *quantized* values. Thus, if the energy levels of a particular atom are E_1, E_2, and E_3 (Fig. 8.1), the atom can never possess an energy between E_1 and E_2 or between E_2 and E_3. The quantum assumption cannot be derived from the classical mechanics of moving objects as applied to moving atoms; it is a completely independent axiom which must be added to classical atomic and molecular theory. Its justification lies, as does that of all other fundamental scientific axioms, in the great power it has given us in the understanding of known behavior and the prediction of new and unsuspected aspects of natural phenomena.

Usually, the student is first introduced to the quantum concept in connection wth atomic spectra. The interaction of radiation with matter to

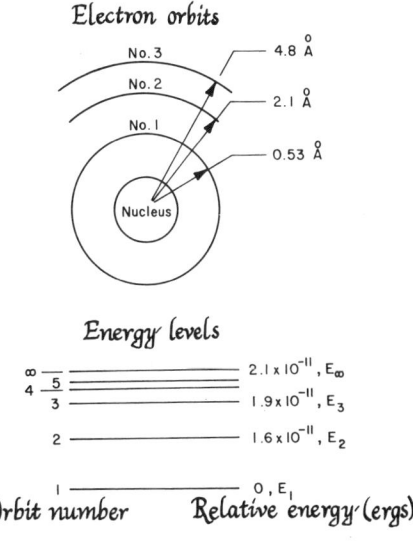

Fig. 8.1. Energy levels in a hydrogen atom.

produce discrete line spectra can be explained in terms of transitions of an electron from one energy level in an atom to another (Fig. 8.1). For a first approximation, we may associate each energy level with an orbit of an electron's motion around the nucleus. For simplicity, we may consider the hydrogen atom, with only a single electron (Fig. 8.1). In its lowest energy state, the ground state, the hydrogen atom has its electron in orbit 1. If radiation of sufficient energy (1.6×10^{-11} erg) is absorbed by the atom, the electron may be boosted to orbit 2. With absorption of 1.9×10^{-11} erg of energy, the electron may jump from the lowest level to orbit 3. There exist an infinite number of orbits, but the energy spacing between successive orbits becomes smaller and smaller as we go up the scale. If an energy slightly greater than 2.1×10^{-11} erg is imparted to the atom, then the electron is actually separated from the hydrogen nucleus, and a positively charged hydrogen ion is left. Hence, the energy difference between orbits 1 and ∞ is named the *ionization energy*. For our purposes, the main point to be noticed in Fig. 8.1 is that there is no energy level between, for example, 0 and 1.6×10^{-11} erg and that, therefore, hydrogen atoms cannot absorb radiation with an energy of, for example, 1.0×10^{-11} erg in each quantum.

In molecules, too, there are different energy levels due to the displacement of an electron from a lower to a higher orbit. The spacings between these energy levels are also of the order of 10^{-11} erg. Furthermore, there are other forms of motion, some of which are not present in atoms (Fig. 8.2), and with each of these motions there is an associated energy.

Two connected atoms in a molecule are joined by a valence bond which may be visualized as a spring. The two atoms, thus, are capable of vibrating toward and away from each other. The frequency of this vibration depends on the energy imparted to the vibrating atoms. In intramolecular vibrational motions, as in electron displacements, the energies cannot be any value but must be only certain discrete values (Fig. 8.2). The spacing between vibrational energy levels, however, is smaller than between

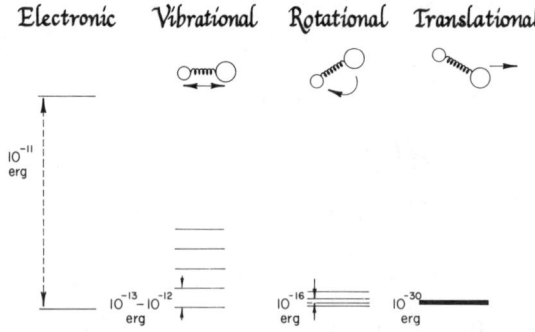

Fig. 8.2. Energy levels in a molecule.

A. Fundamental Assumptions

electronic levels, the former being of the order of 10^{-13}–10^{-12} erg. The vibrational energies that are permissible in any given case depend on the structure of the molecule. In general, the spacing between energy levels is larger the greater is the force of attraction between the vibrating atoms and the smaller are the masses of the atoms.

In addition to vibrational motion, a molecule may be involved in rotational motions which again have only certain discrete energies associated with them (Fig. 8.2). Rotational energies are even smaller than vibrational, being of the order of 10^{-16} erg. The spacing between these levels, in general, depends on the masses of the atoms in the molecule and on the interatomic distances.

Finally, molecules, and atoms too, may move with respect to some reference point in space; that is, they may have translational motion. Once again, quantum theory states that only discrete energies of translation are possible. However, in contrast to other molecular motions, the spacings between translational energy levels are extremely small, being of the order of magnitude of 10^{-30} erg (Fig. 8.2), the actual value depending on the total mass of the molecule and the size of the container in which it is confined. It is undoubtedly clear even from the very schematic drawing of Fig. 8.2 that translational energy levels are so close together, they practically merge into a continuum of energy. In fact, for most practical purposes, they can be treated as a continuum; that is, we may ignore the quantization of translational motion. As a consequence it can be shown that the average energy of translation of a mole of particles depends only on the temperature and is the same for all molecules or atoms:

$$E_{\text{translation}} = \tfrac{3}{2} RT \quad \text{(per mole)}. \tag{8.1}$$

2. Statistical Assumptions

We have outlined the various types of energy levels that may exist in a molecule. In a group of molecules of a single substance, not every molecule has exactly the same energy despite the fact that each molecule has available to it exactly the same energy levels. For example, if we consider the very simple energy level diagram of Fig. 8.3a to be characteristic of some given molecular structure, and if we have a group of 10 molecules of this substance to which we give 10 units of energy, it is exceedingly unlikely that all of the molecules will be in the second energy level at the same time, i.e., in the first level above the ground state (Fig. 8.3b). Rather, the molecules will be distributed among the energy levels in some unsymmetrical fashion consistent with the possession of a total energy of 10 units (Fig. 8.3c), the particular form of the distribution depending on the kind of energy level being considered.

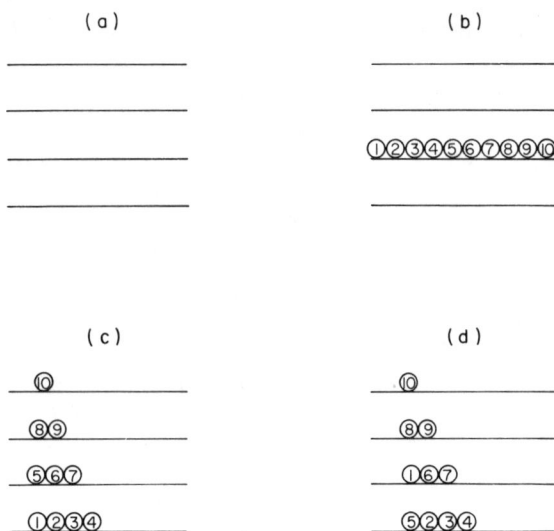

Fig. 8.3. Distribution of molecules among energy levels.

For example, with respect to translational motion along any one direction in space (e.g., along the X axis of the container), a very wide spectrum of energies is possible (Fig. 8.4). A certain number of molecules are in the very lowest level; essentially, an equal number are in the second level, in the third level, etc. In fact, the number of molecules in the milllionth translational level is nearly the same as in the very first. Only in much higher levels, that is, at exceedingly high speeds, does the number of molecules tend to drop off. In other words, the molecules tend to be distributed fairly equally among translational energy levels, except at extremely high energies.

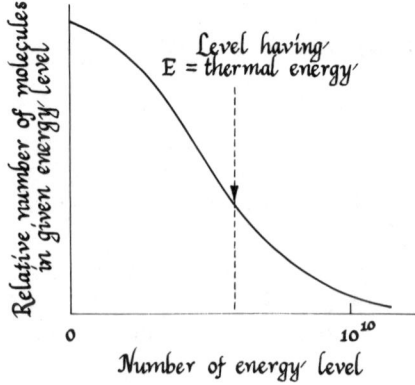

Fig. 8.4. Distribution of molecules among energy levels for translational motion in one direction (along x axis).

A. Fundamental Assumptions

The basic reason for this type of distribution is easy to understand once we are aware of the fact that thermal energy is of the order of 4×10^{-14} erg. This quantity is very large compared to the spacing between translational energy levels (Fig. 8.2). Thus, many molecules can be in high translational states, as well as in low ones. Only for exceedingly high speeds is thermal energy insufficient to supply the required translational energy to more than a few molecules.

In fact, this argument can easily be put into quantitative terms. Suppose some molecules are at equilibrium in their distribution between two levels with energies of E' and E'' (in calories per mole). Let the number of molecules in each level at equilibrium be represented by n' and n'', respectively. Then, the equilibrium constant,

$$K = n''/n', \tag{8.2}$$

may be inserted into Eq. (4.1) to give

$$\Delta G° = -RT \ln(n''/n'). \tag{8.3}$$

From Eq. (3.5) we obtain a replacement for $\Delta G°$:

$$\Delta E° - T\Delta S° = -RT \ln(n''/n'). \tag{8.4}$$

If the two energy levels E' and E'' have no intrinsic preference, one over the other, for the molecules, then the intrinsic probability of the molecules being in level E' is the same as that of being in level E''. Under these circumstances (as we show in Section B) S'' and S', the entropies of the molecules in these levels, are the same, and hence $\Delta S° = 0$. Thus, rearranging Eq. (7.4) to the exponential form we obtain*

$$\frac{n''}{n'} = \exp\left(-\frac{\Delta E°}{RT}\right) = \exp\left[-\frac{(E'' - E')}{RT}\right]. \tag{8.5}$$

The larger the difference $(E'' - E')$, the smaller will be the right-hand side of Eq. (8.5) and hence the smaller the number of molecules n'' in the higher energy level, relative to n'. Figure 8.4 shows in essence a graphical representation of this distribution of molecules among energy levels for translational motion. If we consider energy per molecule (ε) instead of per mole (E) we can modify Eq. (8.5) slightly to read

$$\frac{n''}{n'} = \exp\left[-\frac{(E'' - E')/N}{RT/N}\right] = \exp\left(-\frac{\Delta \varepsilon}{kT}\right), \tag{8.6}$$

where N is Avogardro's number, and $k = R/N$ is called the Boltzmann constant, in recognition of the discoverer of this distribution equation, the

---
* $e^{x(y+z)} = \exp[x(y+z)]$.

Boltzmann equation. Thermal energy is defined as RT (per mole) or kT (per molecule) and thus equals approximately 0.6 kcal/mole or 4×10^{-14} erg molecule at room temperature.

For translational motion $\varepsilon_2 - \varepsilon_1$ between the first and second successive levels is very small, ($\sim 10^{-30}$ erg). Clearly, $\Delta\varepsilon/kT$ is a very small number, and the exponential term on the right-hand side of Eq. (8.6) is essentially unity. Thus, n_2 the number of molecules in the second level is essentially the same as n_1. Only for the billionth energy level does $\varepsilon_{10^9} - \varepsilon_1$ become appreciable compared to kT and only then does n_{10^9}/n_1 begin to drop appreciably below 1.

Likewise, with respect to rotational motion, some molecules occupy the lower energy levels, others the higher. Here again thermal energy (4×10^{-14} erg) is relatively large compared to the spacing between energy levels, (10^{-16} erg), and hence the molecules are distributed not too disproportionately among the energy states.

In contrast, in vibrational motion, where the spacing between levels increases in magnitude to round 10^{-13} erg, most of the molecules, if at room temperature and not excited by illumination, tend to be in the lowest energy level, with only a few in higher energy levels. Such a heavily slanted distribution arises from the fact that in the absence of radiation, again only thermal energy is available to excite molecules. A few molecules may be able to accumulate somewhat more than this quantity of energy at room temperature, but the overwhelming mass of them will be near this average. It is clear from Fig. 8.2 that while 4×10^{-14} erg is enough to put a molecule in a high translational or rotational level, it is insufficient to raise a molecule very high in the vibrational scale. Similar conclusions can be reached from the Boltzmann law [Eq. (8.6)]. Likewise, it is immediately obvious from Fig. 8.2 or from the Boltzmann equation that all the molecules will be in the lowest electronic energy state since thermal energy is several orders of magnitude too small to supply the 10^{-11} erg necessary to excite an electron to an orbit above the ground state.

Except for the electronic energy level, a particular molecule is unlikely to stay in a given energy level very long. Collisions occur constantly between molecules, with accompanying exchanges of energy, and as a result transitions from one energy level to another are constantly in progress, even for a system in equilibrium. For a system in equilibrium containing a large number of molecules, there will be a definite average apportioning of molecules among the different energy levels. Although different molecules may occupy different levels at different times, on the average a particular level will have the same number of tenants. It should be clear, nevertheless, that a particular average distribution can be obtained in many different ways depending on which individual molecules of the

A. Fundamental Assumptions

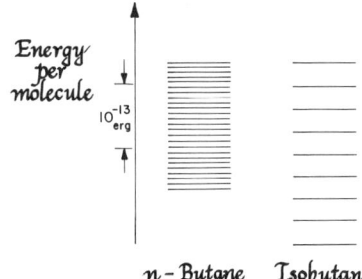

Fig. 8.5. Statistical and energetic factors contributing to molecular stability.

group happen to be residing in each energy level. For example, in connection with the energy levels of Fig. 8.3, a distribution of four molecules in the lowest level, three in the second, two in the third and one in the fourth can be obtained as shown in Fig. 8.3c; it can also be obtained, with one rearrangement, in the alternative fashion shown in Fig. 8.3d; and many other permutations can be written down readily.

Every molecular structure has a definite set of energy levels corresponding to it. For example, if we have a large collection of hydrogens and carbons in the ratio of four carbon atoms to ten hydrogen atoms, these may combine to form molecules of normal butane (**I**) which have a set of levels such as shown in Fig. 8.5. The same atoms combined to form an alternative structure, isobutane (**II**), have a different set of energy levels (Fig. 8.5). The problem that we wish to consider is, which configuration will be preferred by the molecules; that is, which molecular structure, (**I**) or (**II**), is more probable.

```
                                H   H   H
                                |   |   |
                            H — C — C — C — H
                                |   |   |
    H   H   H   H                 H  |   H
    |   |   |   |                    |
H — C — C — C — C — H            H — C — H
    |   |   |   |                    |
    H   H   H   H                    H

      (I)                         (II)
```

Before elaborating on this problem, let us consider an analogous but more familiar probability question. For theoretical or practical purposes we might wish to know which sum is most likely to appear in a single toss of two (perfectly balanced) dice. Without going into any of the theorems of probability, we could answer this question simply by making a list of each of the possible sums that can be obtained (Table 8.1) and then enumerating for each entry all the possible ways in which this number can be thrown.

TABLE 8.1

Statistics of Dice

Sum for single throw	Individual arrangements giving sum		Total ways of obtaining sum
	Die 1	Die 2	
2	1	1	1
3	1	2	2
	2	1	
4	1	3	3
	2	2	
	3	1	
5	1	4	4
	2	3	
	3	2	
	4	1	
6	1	5	5
	2	4	
	3	3	
	4	2	
	5	1	
7	1	6	6
	2	5	
	3	4	
	4	3	
	5	2	
	6	1	
8	2	6	5
	3	5	
	4	4	
	5	3	
	6	2	
9	3	6	4
	4	5	
	5	4	
	6	3	
10	4	6	3
	5	5	
	6	4	
11	5	6	2
	6	5	
12	6	6	1

B. Relationship to Thermodynamic Quantities

For example, there is only one way in which the sum 2 can be thrown, that is, by having 1 appear on both dice. In contrast, there are three different distributions (see Table 8.1) that lead to number 4. If we continue to list the individual arrangements of the two dice that can lead to each sum, we find quickly that the largest number of arrangements can be made for the number 7. Hence the number 7 is the most probable one, the one most likely to appear.

The basic postulate of molecular statistics is similar in nature to that in the throwing of dice. In essence, it says that if we have two different sets of energy levels, starting at the same ground state, one set corresponding to one molecular structure and the second to an alternative structure, then the atoms will tend to arrange themselves into the molecular structure that has available to it the larger number of energy levels. The more energy levels available in a given set to the assembly of molecules, the larger is the number of ways in which the molecules can be distributed among the levels. The larger the number of different ways of distributing the molecules, the more probable is the structure which corresponds to that set of energy levels. For example, the energy levels of normal butane are much more closely spaced than those of isobutane (Fig. 8.5). Consequently, from probability considerations alone, C_4H_{10} is more likely to be in the form of n-butane than isobutane. In general, a more flexible or loosely-arranged structure, e.g., (**I**), has more closely spaced energy levels accessible to it than does a molecular structure with atoms constrained in a tighter arrangement, e.g., (**II**). For that reason, as we shall see shortly, the former is associated with a larger entropy than is the latter.

B. RELATIONSHIP TO THERMODYNAMIC QUANTITIES

The properties of the butane system also point up the one additional factor that we must not forget in applying the reasoning of molecular statistics. Although statistical considerations emphasize the stabilization provided by a high density of energy levels, we must not discard entirely the classical mechanical viewpoint that a system tends toward the state of minimum internal energy. Thus, as shown in Fig. 8.5 the ground state of isobutane is lower in energy than the ground state of n-butane; when each molecule is in a state of rest, the various attractive and repulsive interactions between the four carbon and ten hydrogen atoms leads to a state of lower energy for the isobutane configuration than for normal butane. As a result, there is a pull toward the isobutane structure due to energetic factors that can counteract, in part, the tendency toward the normal butane structure favored by probability effects.

The counteracting effects of energy and probability factors can be demonstrated even on a microscopic scale. If, for example, 500 white balls and 500 red balls are placed in a box and shaken vigorously, then the chances are very high that the bottom layer will have approximately as many white balls as red ones. If, however, the red balls have small pieces of iron in them and a large magnet is placed below the bottom of the box, then after vigorous shaking we will find many more red balls than white balls in the bottom layer. Thus the "striving" toward a distribution of maximum probability is counteracted by a tendency to reach to state of lower energy.*

It becomes clear, then, that the net result will depend on the relative magnitude of these two effects, which, for molecular systems, we wish to express in quantitative terms.

Since the net result of the counterbalancing of the energetic and statistical factors is a measure of the tendency toward formation of one molecular structure from some other, the resultant quantity obviously has the characteristics of the chemical potential. Thus, it should seem reasonable that statistical energetics leads us to the relation

$$\Delta G = \text{(Change in ground-state energy)} - \text{(Change in probability of system)}. \tag{8.7}$$

We need only to recall that a transformation can occur spontaneously when ΔG is negative; ΔG can be negative, according to Eq. (8.7), under either of two circumstances: (1) if the change is energy in the ground state ΔE_0 is negative (that is, the energy drops) or (2) if the probability of the configurations is greater at the end than at the outset. A negative sign must precede the probability term in Eq. (8.7), for an increase in probability favors a reaction, that is, should lead to a *negative* value for ΔG.

More specifically, statistical thermodynamics leads to an equation of the following type in place of Eq. (8.7)

$$\Delta G = \Delta E_0 - RT \ln Z, \tag{8.8}$$

where Z is a measure of the relative probability of the configuration of the products of the reaction as compared to the reactants.

It is instructive at this point to recall Eq. (3.5),

$$\Delta G = \Delta E - T\Delta S. \tag{8.9}$$

The first term on the right-hand side of Eqs. (8.8) and (8.9) are not exactly the same since in Eq. (8.8) ΔE_0 is the change in internal energy when the molecules are essentially without any motion, that is, at the absolute zero of temperature, whereas in Eq. (8.9) ΔE refers to the change in

---

* Likewise in weighted dice, the intrinsic probability factor is counterbalanced by the tendency of the dice to attain minimum gravitational energy.

internal energy at the temperature of the reaction. However, in our present qualitative discussion this distinction may be neglected. In consequence, we see from a comparison of the second terms that

$$\Delta S = R \ln Z. \tag{8.10}$$

In other words, from the molecular-statistical viewpoint, it is the entropy change which is related to the relative probability. An increase in probability of the system, thus, is the molecular basis for an increase in entropy. From the atomic theory, therefore, we are able to obtain a mechanical interpretation for the abstract concept of entropy originally introduced in classical energetics.

C. SOME APPLICATIONS

It remains for us to consider briefly what molecular structural factors lead to increased probabilities, increased entropies, and hence negative ΔG's. We should recall that these probabilities are determined by the density in energy levels, that is, by the closeness of their spacing. The energy levels, in turn, are determined by molecular motions. In a general way, we can say that the greater the freedom of motion, the closer are the energy levels; on the other hand, any restraints on the movement of the molecules or any interference with the freedom of motion within the molecule increases the spacing between energy levels. The significance of these general statements can be made more evident if we consider a few examples.

A particularly simple case is the melting of ice:

$$H_2O(s) \rightleftharpoons H_2O(l). \tag{8.11}$$

As we are all aware, this process proceeds spontaneously to the right if the temperature is a fraction of a degree above 0°C. From energy considerations alone, such a transformation would be an anomaly since it requires an input of energy, approximately 1400 cal/mole (Fig. 8.6). From a molecular-statistical viewpoint, however, it is immediately evident that H_2O molecules in water have much greater freedom of motion than they do in ice. Consequently the factor Z, the measure of the relative probability of the liquid configuration compared to that of the crystalline form, is greater than unity, and hence ΔS is positive. It is this increase in entropy (Fig. 8.6) which (when multiplied by T) more than compensates for ΔE and hence leads [Eq. (8.9)] to a drop in free energy, as we would expect for a spontaneous process.

For water in its various states—gas, liquid, and solid—the corresponding entropies have been determined experimentally (Table 8.2). As we would expect, the entropy increases in the order solid < liquid < gas,

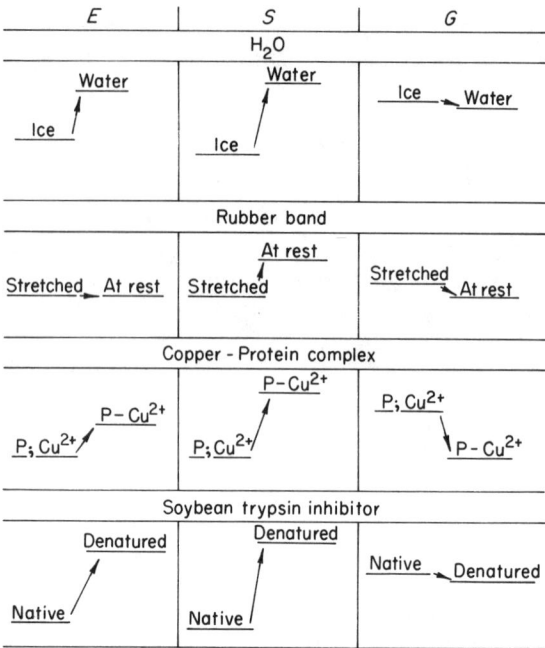

Fig. 8.6. Some processes that occur spontaneously despite unfavorable changes in internal energy.

paralleling the increase in freedom of motion of the molecules. Similarly the entropy increases with increasing softness of a metal (Table 8.2), i.e., with increasing ease of separation of atoms even in the solid state. Likewise entropy increases with increasing molecular weight, with increasing complexity, and with increasing flexibility of a molecule (Table 8.2). Thus, even for monoatomic gases, which have no rotational or vibrational motions (Fig. 8.2), the density of translational energy levels increases, because the spacing between these levels decreases with increasing molecular weight. Consequently the entropy of argon is well above that for helium. In the polyatomic molecules, energy levels appear due to vibrations and rotations, and the number of levels increases with the number of atoms in the molecule and with their weight. Thus, the entropy increases in a parallel fashion (see Table 8.2).

In principle, it should be possible to compute the absolute entropy of a substance if we know full details about its molecular structure and motions. In practice, these calculations are difficult and often intractable. A simple example may illustrate one feature of such quantitative calculations.

We start with Eq. (8.10). For molecular-statistical calculations, the relative probability Z is defined in a different manner than in common

C. Some Applications

TABLE 8.2

Entropy and Molecular Disorder

Entropy and Physical State: H_2O at 0°C	
State	Entropy[a]
Ice	9.8
Liquid	15
Vapor (1 atm)	45

Entropy and Hardness		
Element	Entropy[a]	Hardness
C (diamond)	0.6	10
B	1.4	9.5
Si	4.5	7
Cu	8.0	2.7
Ca	10.0	1.5
Na	12.2	0.5

Entropy, Molecular Complexity, and Molecular Weight: Gaseous Molecules at 1 Atm and 25°C

	Molecule	Molecular weight	Entropy[a]
Monatomic			
	H	1	27
	He	4	30
	A	40	37
	Xe	131	41
Polyatomic			
	H_2	2	31
	N_2	28	46
	I_2	254	63
	H_2O	18	45
	H_2S	34	49
	CO_2	44	51
	NH_3	17	46
	CH_4	16	44
	CCl_4	154	74
Flexibility			
	iso–C_4H_{10} (II)	58	70
	n–C_4H_{10} (I)	58	74

[a] Calories per mole per degree

statistical calculations, where Z varies only between zero and unity. In molecular statistics, the probability Z is measured by the total number of different arrangements into which a molecular system may be placed. For example, at temperatures approaching absolute zero, the molecules in a crystalline substance such as solid HCl lose their various twistings and vibrations, drop to their lowest available energy level and settle down into a perfect crystalline pattern, represented schematically by

$$\text{HCl HCl HCl HCl} \ldots.$$

Under these circumstances there exists *only one* possible arrangement of the molecules in the crystal. Therefore Z is unity, and it follows from Eq. (8.10) that the entropy of crystalline HCl at 0 K is zero.

The molecules of CO in the crystalline state also lose their molecular vibrations and twistings as $T \to 0$ K, and they too settle down into their lowest energy level. In this case, however, it turns out that the crystal does not attain a perfect array. Because the carbon and oxygen atoms are nearly equal in size, as the temperature approaches 0 K, the molecules can settle into the lattice lined up either as CO or as OC. Thus, the crystalline pattern may be represented schematically as

$$\text{CO CO OC CO} \ldots.$$

Obviously there are two possible arrangements for each CO molecule; the value of Z becomes 2. We would compute, therefore, from Eq. (8.10) an entropy of $R \ln 2 = 1.38$ (cal/mole/deg) for crystalline carbon monoxide at 0 K. The calculated value agrees closely with the experimentally measured value of 1.1 cal/mole/deg.

In crystalline ice at 0 K there is a different type of disorder. All of the oxygen atoms of the H_2O molecules in the crystal stay in fixed positions. However, when we consider hydrogen-bonding in the lattice (shown around an oxygen atom in two-dimensional projection **III**) we recognize

```
            O
            |
            H
            ·
            ·
            ·       |← 2.76 Å →|
    O—H ····O—H ···· O
            |   ↖0.99Å
            H
            ·
            ·
            ·
            O
```

(III)

that in any O—H—O bond, the hydrogen atom could occupy either of two positions, e.g.,

$$\text{O—H}\cdots\text{O} \quad \text{or} \quad \text{O}\cdots\text{H—O}.$$

C. Some Applications

If each hydrogen atom can move independently, then since there are 2 moles of hydrogen per mole of water, the entropy [from Eq. (8.10)] becomes $2R \ln 2$, or approximately 2.7 cal/mole deg. The experimental value at 0 K, however, is known to be much lower (0.82 cal/mole deg). Clearly this form of theoretical calculation must be an oversimplified one. On further inspection, however, we recognize that each H could not really move independently. For example, if that on the east side of the central oxygen atom in **III** moves to the right, a negative charge is left on the central oxygen and a positive charge is placed on the east oxygen. However, such a separation of charge would require an input of energy, and hence is not likely to occur. Thus, the east hydrogen atom can move to the east oxygen only when (for example) the north hydrogen atom moves down to the central oxygen so that the latter carries no net charge. When these electrostatic constraints on the number of arrangements are fully worked out, the number of possible positions of the hydrogen atoms is reduced substantially and the corrected theoretical computation leads to an entropy at 0 K of 0.80 cal/mole deg, which is in good agreement with experimental observation.

In the example of the transition of ice to water described earlier (Fig. 8.6), the direction of the entropy change is immediately obvious from visible macroscopic properties of the substance. Even more interesting are those cases in which no simple macroscopic observation reflects the molecular state and, as a consequence, a molecular statistical analysis provides some insight into configurational changes at the molecular level.

From common experience we know that a stretched rubber band can snap back spontaneously to the rest position. Clearly then ΔG must be negative for this process (Fig. 8.6);

$$\text{rubber band (stretched)} \rightleftharpoons \text{rubber band (at rest)}; \quad \Delta G < 0. \tag{8.12}$$

Experimental measurements show, however, that ΔE is nearly zero. It follows then [Eq. (8.9)] that ΔS must be positive. In other words, the entropy of the resting rubber band is greater than that of the stretched state. We are thus led to the conclusion, which could not otherwise have been anticipated, that the molecules in rubber bands possess greater freedom of motion in the resting state than in the stretched. Such a situation could be obtained if the very long molecules from which rubber is made were highly oriented (Fig. 8.7a) in the stretched state but more disorganized and randomly arranged in the resting state (Fig. 8.7b). Freedom of motion would be hampered in (a) because of the essentially crystalline configuration. Consequently (a) would have a lower entropy than (b). From probability considerations, therefore, the transition from stretched to resting state would be favored. Since ΔE is essentially zero, ΔS alone is the determining factor in making ΔG negative for the snapback.

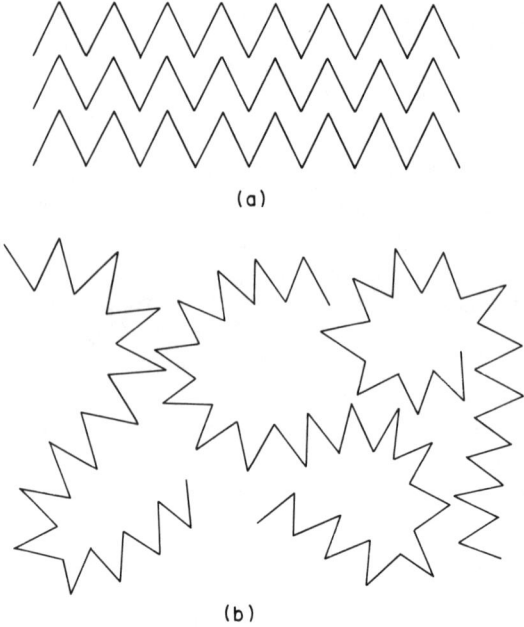

Fig. 8.7. Conformations of macromolecules in rubber: (a) ordered and (b) random.

Turning to a chemical rather than physical transformation, we might examine again one of the processes mentioned earlier in passing (Fig. 2.1), the formation of a complex between Cu^{2+} and protein P:

$$P + Cu^{2+} \rightleftharpoons P\text{--}Cu^{2+}. \tag{8.13}$$

In this case the process proceeds spontaneously to the right despite the fact that the internal energy of the complex P–Cu^{2+} is 3 kcal/mole greater than that of separated P and Cu^{2+} (Fig. 8.6). Experimental measurements show that ΔS accompanying reaction (8.13) is a large positive number. At first glance such a result would be most surprising, for we would presume that the free components P and Cu^{2+} would have greater freedom of motion than the complex P–Cu^{2+} in which the components are bound together and cannot move about independently. Nevertheless if ΔS is known to be positive, there must be some increase in freedom of motion when reaction (8.13) proceeds to the right. The most likely molecular interpretation of this puzzle draws attention to one feature of the reaction which Eq. (8.13) does not indicate explicitly—that the participants are all dissolved in aqueous solution. The molecules of P and Cu^{2+} must be hydrated; that is, water molecules must be bound rigidly to these particles.

C. Some Applications

If, as we would reasonably expect, some of this "frozen" water is released when P and Cu^{2+} combine to form a complex, then more molecules are given increased freedom of motion than are restrained in their movement due to formation of $P-Cu^{2+}$. Thus, in place of Eq. (8.13) we really should write

$$P(H_2O)_x + Cu(H_2O)_y^{2+} \rightleftharpoons P-Cu^{2+} + (x+y)H_2O, \qquad (8.14)$$

where x represents the number of H_2O molecules released by the protein, and y the number released by the copper when the complex is formed. This equation indicates explicitly that "frozen" water molecules are released simultaneously with the formation of the complex, and hence we can understand why the entropy may increase during this reaction.

Finally we might cite another chemical transformation of biological interest, the denaturation of proteins. One of the best examples for thermodynamic analysis is the protein that inhibits the proteolytic enzyme trypsin, called soybean trypsin inhibitor. Thermodynamic data[*] for the denaturation of soybean trypsin inhibitor,

$$\text{protein (native)} \rightleftharpoons \text{protein (denatured)}, \qquad (8.15)$$

show that there is an enormous increase in internal energy (57 kcal/mole) on formation of the denatured form. Nevertheless, at temperatures near 50°C, the denaturation proceeds spontaneously, that is, ΔG is negative. Again, it follows from Eq. (8.9) that ΔS must be a large positive number. Denaturation must be accompanied, therefore, by a large increase in freedom of motion of the molecular system. This increased freedom is generally assigned to the protein molecule. In the native state, a protein must be a highly oriented system (Fig. 8.8); if in the denatured form it becomes unfolded in part, it is clear that the freedom of motion within the molecule will be greatly increased. The large positive ΔS thus suggests that the protein molecule becomes disoriented during denaturation.

It is, in fact, possible to use Eq. (8.10) to estimate the entropy change to be expected in protein unfolding. We assume that in the native folded form each amino acid residue has a fixed inflexible position (see Fig. 8.8). In the unfolded state, each residue that is freed from constraints acquires additional possible orientations. Since three interatomic bonds are contributed by each residue in a polypeptide chain we shall make the very simplified assumptions that there are three points of flexibility per residue and that at each such point there are two possible orientations (of equal energy).[†]

---

[*] M. Kunitz, *J. Gen. Physiol.* **32**, 257 (1948).

[†] W. Kauzmann *in* "The Mechanism of Enzyme Action," (W. D. McElroy, and B. Glass, eds.), pp. 70–120. Johns Hopkins Press, Baltimore, 1954.

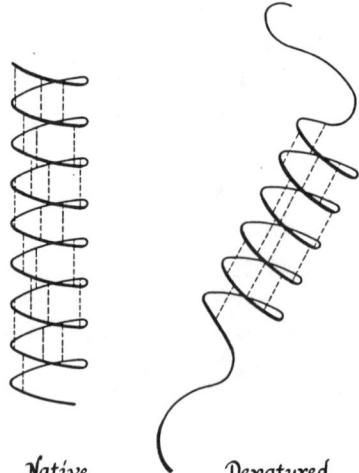

Fig. 8.8. A simplified model for protein denaturation.

Native / Denatured

Then the total number of new possible orientations contributed by each residue in the unfolded state is 2^3. Therefore,

$$\Delta S = n(R \ln 2^3) = 4.1\, n \text{ (cal/mole deg)} \tag{8.16}$$

for the "conformational entropy" of unfolding of a protein molecule of n residues. At 25°C this contributes $-T\Delta S = -1200n$ calories to ΔG for unfolding. This is a large entropic contribution by each residue to the thermodynamic stability of the denatured state. It seems likely that not every residue in the protein is completely freed from constraints upon denaturation.

These examples illustrate some of the information on molecular characteristics which we can obtain from a statistical thermodynamic approach to material transformations. From the molecular-statistical viewpoint, in contrast to the classical approach, we can deduce definite information on the relative freedom of motion of molecular systems. A detailed assignment of this freedom to particular parts of the molecular system is often less certain, however. For example, in the examples described with proteins, highly complex molecules, it is quite possible that the entropy of denaturation is due to changes in bound water rather than to rearrangements in the structural framework of the protein molecule. Nevertheless, the number of alternative explanations available in a given situation is small. Each of them suggests new types of experiments. It is in the ability to interrelate seemingly diverse experimental approaches that the special usefulness of molecular statistics lies.

9

Energetics of Enzyme Kinetics

Catalytic effects of enzymes have been interpreted in the terminology of different conceptual models. The most universally used one at present is transition-state theory.

The central concept of transition-state theory is that reactants are in equilibrium with a transition-state species, which then proceeds to create products. For example, for the simple chemical displacement reaction

$$CH_3I + H_2O \rightarrow CH_3OH + HI, \qquad (9.1)$$

we visualize the following molecular path through a transition state,

$$\underset{H}{\overset{H}{\underset{\diagdown}{O}}} + H \blacktriangleright \underset{H}{\overset{H}{\underset{\diagdown}{C}}} - I \; \underset{\longleftarrow}{\overset{equilibrium}{\longrightarrow}} \; \left[H_2O \cdots \underset{H}{\overset{H}{\underset{\diagdown}{C}}} \cdots I \right]^{\ddagger} \longrightarrow \underset{H}{\overset{H}{\underset{\diagdown}{O}}} - \underset{H}{\overset{H}{\underset{\diagdown}{C}}} \blacktriangleleft H + H^+ + I^- \quad (9.2)$$

$$\text{transition state}$$

95

In general, for a unimolecular reaction with one reactant S, we may write

$$S \rightleftharpoons S^\ddagger \rightarrow P, \quad (9.3)$$

and for a bimolecular reaction, with two reactants S and N,

$$S + N \rightleftharpoons \underset{\text{intermediate complex}}{S \cdot N} \rightleftharpoons \underset{\text{transition state}}{[S \cdot N]^\ddagger} \longrightarrow P \quad (9.4)$$

A. CONCENTRATION PROFILES

The rate of production of products P is proportional to the concentration of the transition-state species from which the products arise. To alter the rate of a reaction we must change the concentration of the transition-state species. Thus, it is useful, in order to understand catalysis or inhibition, to examine the relative concentrations of participating species. In terms of a chart, what we need is a *concentration profile*.

For example, for a unimolecular reaction, relative concentrations can be represented on a chart such as Fig. 9.1. (A logarithmic scale is used because $(S^\ddagger)$ and (S) may differ by a factor as large as 10^{15}, which would be difficult to represent on a linear concentration scale). If we can raise the

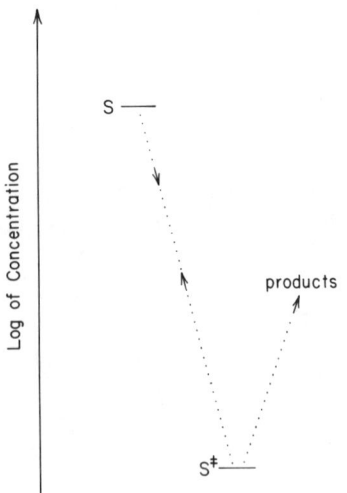

Fig. 9.1. Concentration profile for a straightforward (noncatalyzed) unimolecular reaction.

A. Concentration Profiles

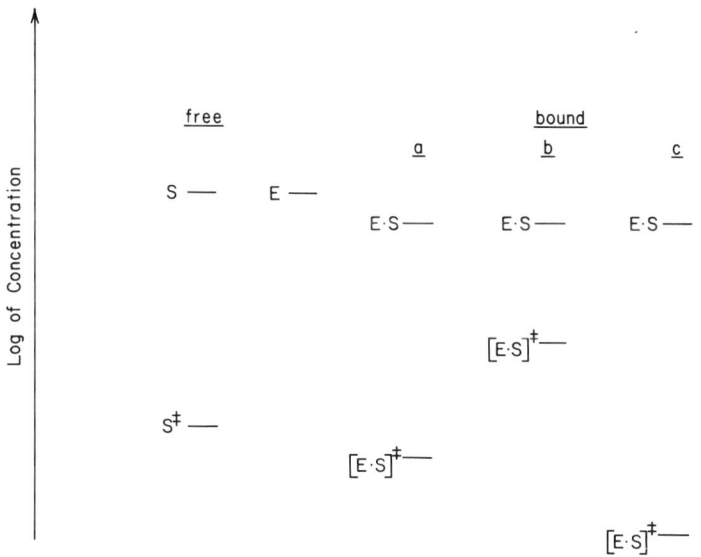

Fig. 9.2. Concentration profile for an enzyme-catalyzed reaction when only a single reactant S is bound by the enzyme E.

level of $S^{\ddagger}$ (i.e., increase its concentration, for example, by increasing the temperature), then the rate of production of products will increase.

Enzymes can also increase rates of reaction. To do so they must increase the concentration of activated, transition-state species above that obtained in the absence of enzyme. A set of possible concentration profiles when only a single reactant is bound by the enzyme E is shown in Fig. 9.2. In the presence of E, the substrate may be bound (in the *nonactivated* state) to produce the species E·S. The concentration of the latter will depend on the magnitude of the binding constant K_{ES}, i.e.,

$$E + S \rightleftharpoons ES; \qquad K_{ES} = (E \cdot S)/(E)(S). \tag{9.5}$$

Insofar as the focal question, the rate of production of products, is concerned, however, the crucial issue is what is the concentration of $[E \cdot S]^{\ddagger}$?

To approach this problem, let us also write equilibrium constants for the formation of the respective transition-state species:

$$K_S^{\ddagger} = (S^{\ddagger})/(S), \tag{9.6}$$

$$K_{ES}^{\ddagger} = ([E \cdot S]^{\ddagger})/(E \cdot S). \tag{9.7}$$

We then find that if the equilibrium constant for the formation of $[E \cdot S]^{\ddagger}$ from $E \cdot S$ is the same as that for the formation of $S^{\ddagger}$ from S outside the enzyme–substrate complex (column a in Fig. 9.2) then *no acceleration* in rate will be effected by the supposed enzyme. For under these circumstances

$$K_S^\ddagger = \frac{(S^\ddagger)}{(S)} = K_{ES}^\ddagger = \frac{([E \cdot S]^\ddagger)}{(E \cdot S)} = K^\ddagger, \tag{9.8}$$

and the ratio of concentration of all forms of transition states to all forms of ground state becomes

$$\frac{(S^\ddagger) + ([E \cdot S]^\ddagger)}{(S) + (E \cdot S)} = \frac{K^\ddagger(S) + K^\ddagger(E \cdot S)}{(S) + (E \cdot S)} = K^\ddagger = K_S^\ddagger. \tag{9.9}$$

In other words, the proportion of activated transition-state species has not been changed by the addition of supposed enzyme. In essence, since the *fraction* of S that becomes activated is not changed by the presence of the macromolecule there can be no change in rate.

It follows, therefore, that in a single-substrate reaction, a catalytically effective enzyme must have a $K_{ES}^\ddagger$ that is larger than $K_S^\ddagger$ in the absence of macromolecule. Only then does it follow (from algebraic manipulations analogous to those in Eqs. (9.6)–(9.9) that the fraction of transition-state species in the presence of enzyme is greater than in its absence.* In a single-substrate reaction, a catalyst must provide an environment conducive to the formation of activated species if the presumed catalyst is to effect an acceleration in rate of reaction.

If the enzyme does favor formation of the activated species, then the concentration of the latter depends not only on $K_{ES}^\ddagger$ but also on the association constant K_{ES} for the formation of $E \cdot S$ [see Eq. (9.5)]. Algebraic analysis leads to the following expression for the fraction $\alpha^\ddagger$ of activated molecules

$$\alpha^\ddagger = \frac{([E \cdot S]^\ddagger) + (S^\ddagger)}{(E \cdot S) + (S)} = \frac{K_{ES}^\ddagger K_{ES}(E) + K_S^\ddagger}{K_{ES}(E) + 1}. \tag{9.10}$$

Thus, the fraction of molecules activated (and, hence, the rate of reaction) depends on the concentration of enzyme (E) and its affinity for substrate, as expressed in K_{ES}. If $K_{ES}^\ddagger$ is greater than $K_S^\ddagger$ and is independent of K_{ES}, then the stronger the binding of S, the more accelerated would be its rate of reaction.† It is also evident from Eq. (9.10) and Figure 9.2 that at very

---

* I. M. Klotz, *J. Chem. Educ.* **53**, 159–160 (1976).

† If $K_{ES}^\ddagger$ is smaller than $K_S^\ddagger$ then the supposed enzyme *decreases* the rate of reaction (column c in Fig. 9.2), the degree of attenuation (or nonproductive binding) depending on the strength of binding and the concentration of E.

A. Concentration Profiles

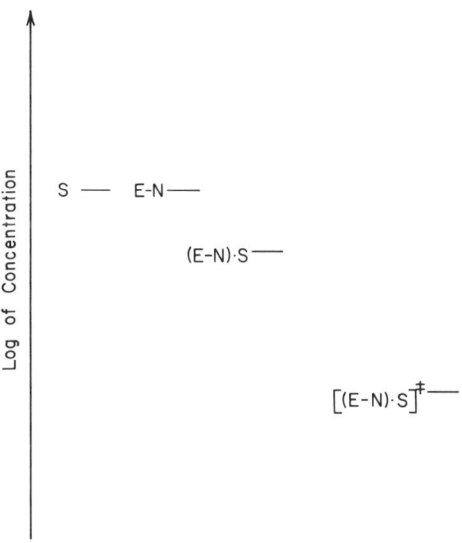

Fig. 9.3. Concentration profile for an enzyme-catalyzed reaction in which one of two reactant species is a specific functional group linked within the enzyme.

high (E) or as K_{ES} becomes increasingly large, $\alpha^{\ddagger}$ tends towards a limit, and the velocity of the reaction will attain a plateau value.

Extending our concentration profile analysis to a two-substrate reaction, we must distinguish between different modes of participation of the macromolecule. Figure 9.3 illustrates concentration relations in a reaction in which the primary substrate S must interact with a second substrate N (e.g., a nucleophile). In the absence of macromolecule, S and N form collision pairs N·S, whose concentration will usually be very small since net attractive forces between these species in solution are usually negligible. The collision pair will then establish an equilibrium concentration of activated complex $[N \cdot S]^{\ddagger}$.

If the enzyme has the second participant covalently linked to it, as E–N denotes, and if this macromolecule has a significant affinity for S, then an appreciable concentration of ((E–S)·S) will appear. In particular

$$((E-N) \cdot S) > (N \cdot S), \tag{9.11}$$

the actual magnitude of ((E–N)·S) depending on the magnitude of the binding constant $K_{(E-N) \cdot S}$. Consequently, even if the equilibrium constant $K^{\ddagger}_{(E-N) \cdot S}$ for the formation of $[(E-S) \cdot S]^{\ddagger}$ is the same as $K^{\ddagger}_{N \cdot S}$ in the absence of the enzyme, the concentration of activated species is greater in the presence of the macromolecule, i.e.,

$$([(E-N) \cdot S]^{\ddagger}) > ([N \cdot S]^{\ddagger}). \tag{9.12}$$

Thus, the velocity of the transformation is increased in the presence of a binding macromolecule even if the fraction of enzyme-bound substrate that is activated to the transition state is unchanged, because the (E–N)·S concentration is much greater than is that of (N·S). If the enzyme in addition provides an environment conducive to the existence of activated species, then the velocity will be increased even further.

It is possible also to have a two-substrate reaction in which neither subtrate is covalently linked to the macromolecule. In that case, the concentration profile would have an appearance similar to Fig. (9.3) but with a bond between E and N absent. The predecessor of the activated complex would be E·N·S. Thus, the comments made with respect to (E–S)·S are valid for E·N·S, but we must recognize one complicating feature in the equilibrium association generating the latter. Since E·N and E·S can also be formed, in addition to E·N·S, increasing concentrations of E relative to N and S will favor formation of the substrate-separated complexes E·N and E·S and decrease the concentration of the two-participant complex E·N·S. Thus, with increasing enzyme concentration the acceleration in velocity may reach a maximum and thereafter actually decline.

B. ACTIVATION ENERGY DIAGRAMS

Since we have defined an equilibrium constant for the formation of the transition-state species [e.g., Eqs. (9.6) and (9.7)], we can also calculate a standard free energy change $\Delta G^{\circ\ddagger}$ for the activation process [compare with Eq. (4.1)]:

$$\Delta G^{\circ\ddagger} = -RT \ln K^{\ddagger}. \tag{9.13}$$

Such $\Delta G^{\circ\ddagger}$ values provide the information necessary for the construction of a free energy diagram, such as that illustrated in Fig. 9.4.

To an inexperienced reader the diagram implies that the catalytic effect of the enzyme E on substrate S is due to the fact that the energy level $\Delta G^{\circ\ddagger}$ for the enzyme-activated complex $[E \cdot S]^{\ddagger}$ is below that for the corresponding nonenzymatic species $S^{\ddagger}$. The very use of the letter E predisposes as to presume a catalytic effect on a reaction rate. In fact, if E' were a molecule that combined strongly with S but slowed down the chemical transformation (i.e., if E' were an inhibitor) the free energy level of E'·S would still be below that of S and hence that of $[E' \cdot S]^{\ddagger}$ could also lie below that of $S^{\ddagger}$ (see Fig. 9.4). Indeed, no acceleration in rate occurs (in the simplest, single-substrate reaction) unless $\Delta G^{\circ\ddagger}_{E \cdot S}$ for $E \cdot S \rightarrow (E \cdot S)^{\ddagger}$ is

B. Activation Energy Diagrams

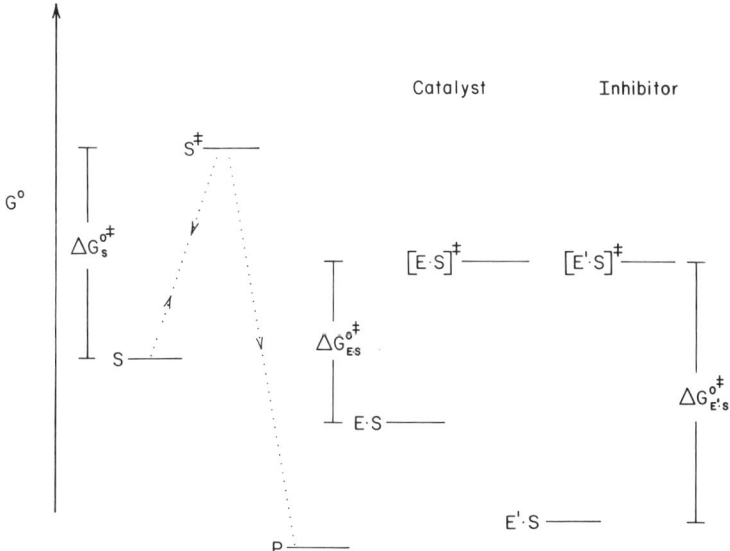

Fig. 9.4. Activation energy diagram showing free energies of activation of different transitions.

smaller than $\Delta G_S^{o\ddagger}$ for $S \to S^{\ddagger}$:

$$\Delta G_{E \cdot S}^{o\ddagger} < \Delta G_S^{o\ddagger}. \tag{9.14}$$

Only under these circumstances will $K_{ES}^{\ddagger}$ be larger then $K_S^{\ddagger}$. Only under these circumstances will the fraction of activated molecules in the transition state be greater in the presence of the enzyme.

In a complementary direction, the reason E' acts as an inhibitor is that

$$\Delta G_{E' \cdot S}^{o\ddagger} > \Delta G_S^{o\ddagger}. \tag{9.15}$$

Thus, even though $\Delta G_{E' \cdot S}^{o\ddagger}$ is below $\Delta G_S^{o\ddagger}$ (see Fig. 9.4) the value of $K_{E' \cdot S}^{\ddagger}$ is smaller than $K_S^{\ddagger}$. Consequently, when S is bound to form E'·S, the fraction of molecules that exist in the activated, transition state is *lowered* rather than increased, and hence the rate of reaction is decreased. The height of an activated species [ ]‡ in a free energy level diagram is not the determining feature in setting the rate; the $\Delta G^{o\ddagger}$ is the controlling quantity.

In summary, to increase a rate of reaction, a catalytic macromolecule must increase the concentration of transition-state species. To achieve this the macromolecule must, as a minimum, bind substrates. Thereafter, if the macromolecule provides an environment facilitating activation, further accelerations in rates of reaction will occur. A concentration profile illustrates the interplay between interrelated species and reveals explicitly the concentrations of activated molecules in the transition state, which determine the reaction rate. A free energy diagram places the interrelated species in energetic order from which transition energies for different steps can be visualized.

EXERCISES

1. For the reaction of Eq. (9.1), $\Delta G^{o\ddagger}$ at 25°C is 27,200 cal/mole.
   (a) Find $K^{\ddagger}$.
   Answer: 1.6×10^{-20}.
   (b) Find the concentration of the transition-state species when $[H_2O] = [CH_3I] = 1$ M.
   Answer: 10^{-20} M.

10

Ligand–Receptor Interactions

Except when radiation participates, all biological activities involve contact interactions between constituents. At the molecular level, if one of the participants is smaller than its complementary partner, the former is usually designated the "ligand" and the latter the "receptor." Thus, in an enzyme–substrate complex, the substrate is the ligand, and the enzyme is the receptor. In immunological interactions the ligand is the hapten or antigen, and the receptor is the immunoglobin. Neurotransmitters or hormones are effector ligands when they are bound to receptor sites at synapses or on cell membranes.

The formation of a ligand–receptor complex is generally the first step in a biological response to an effector. For this first step there are three general features that need to be elucidated at the outset: how many ligands are bound by the receptor entity; how strongly are the ligands held; what is the molecular nature of the forces of interaction? A physicochemical analysis based on thermodynamic principles provides the framework for answering the first two questions and, with some additional molecular information, gives some insights into the third question.

A. MULTIPLE EQUILIBRIA

We shall be considering interactions that are reversible, i.e., interactions in which the ligand and receptor are in equilibrium. Many receptors contain more than one site for binding a ligand. For example, the immunoglobulin IgG can bind two moles of hapten per mole antibody, hemoglobin can bind 4 moles of O_2, aspartate transcarbamoylase can bind six moles of carbamyl phosphate, human serum albumin can hold tens of moles of a wide range of small molecules. Consequently we shall formulate equations for the equilibria which recognize these multiple stoichiometries.

1. Stoichiometric Formulation

a. General statement. There are two distinct approaches to the derivation of equations for multiple equilibria. The first is a purely thermodynamic one, which does not even have to introduce the molecular concept of sites on the receptor. It starts merely by taking cognizance of the multiple stoichiometry seen in binding. Thus if P represents the receptor (generally a protein) and A the ligand, we may write

$$P + A \rightleftharpoons PA$$
$$PA + A \rightleftharpoons PA_2$$
$$\vdots \quad \vdots \quad \vdots$$

For each equilibrium we define a *stoichiometric equilibrium constant*, suitably indexed to distinguish the steps:

$$P + A \rightleftharpoons PA; \quad K_1 = \frac{(PA)}{(P)(A)}; \quad (PA) = K_1(P)(A),$$

$$PA + A \rightleftharpoons PA_2; \quad K_2 = \frac{(PA_2)}{(PA)(A)}; \quad (PA_2) = K_2(PA)(A) = K_1 K_2 (P)(A)^2,$$

$$\vdots$$

$$PA_{i-1} + A \rightleftharpoons PA_i; \quad K_i = \frac{(PA_i)}{(PA_{i-1})(A)}; \quad (PA_i) = (K_1 K_2 \cdots K_i)(P)(A)^i,$$

$$\vdots$$

$$PA_{n-1} + A \rightleftharpoons PA_n; \quad K_n = \frac{(PA_n)}{(PA_{n-1})(A)}; \quad (PA_n) = (K_1 K_2 \cdots K_n)(P)(A)^n.$$

(10.1)

A. Multiple Equilibria

At some high concentration of A (which may not always be attainable experimentally) the receptor will be saturated. The total moles of A bound at saturation is designated as n.

Below saturation the moles of bound A will vary with the concentration of nonbound, free A. The moles of bound A may be related to the concentrations of individual ligand–receptor complexes (PA_i) listed in Eq. (10.1) by the following arguments. Each mole of (PA_1) has one mole of bound A; each mole of (PA_2) contributes two moles of bound A; each mole of (PA_i) contributes i moles of bound A. Consequently,

$$B = \frac{\text{moles of bound } A}{\text{total moles protein}}$$

$$= \frac{(PA_1) + 2(PA_2) + \cdots + i(PA_i) + \cdots + n(PA_n)}{(P) + (PA_1) + (PA_2) + \cdots + (PA_i) + \cdots + (PA_n)}. \quad (10.2)$$

In the denominator of Eq. (10.2) each term has the coefficient 1 since each ligand–receptor species contributes 1 mole of protein. In the denominator is also the concentration (P) since some of the receptor has no ligand attached.

In the right-hand column of Eq. (10.1) are expressions for concentrations of each ligand–receptor species in terms of a product of stoichiometric equilibrium constants and the concentrations (P) and (A). We can now use these to convert Eq. (10.2) into the form

$$B = \frac{K_1(P)(A) + 2K_1K_2(P)(A)^2 + \cdots + i(K_1K_2\cdots K_i)(P)(A)^i + \cdots + n(K_1K_2\cdots K_n)(P)(A)^n}{(P) + K_1(P)(A) + K_1K_2(P)(A)^2 + \cdots + (K_1K_2\cdots K_i)(P)(A)^i + \cdots + (K_1K_2\cdots K_i)(P)(A)^n}. \quad (10.3)$$

It is apparent that since every term in the numerator contains the factor (P) and every term in the denominator contains this factor, then it may be cancelled out, and hence Eq. (10.3) becomes

$$B = \frac{K_1(A) + 2K_1K_2(A)^2 + \cdots + i(K_1K_2\cdots K_i)(A)^i + \cdots + n(K_1K_2\cdots K_n)(A)^n}{1 + K_1(A) + K_1K_2(A)^2 + \cdots + (K_1K_2\cdots K_i)(A)^i + \cdots + (K_1K_2\cdots K_n)(A)^n}. \quad (10.4)$$

Let us digress for a moment to note a relationship that will be very useful in future algebraic manipulations. We can represent the long series in the denominator of Eq. (10.4) by the symbol Z. Notice that if each term in this series Z is differentiated with respect to (A) and then multiplied by (A) we can generate each of the terms in the numerator of Eq. (10.4). In algebraic symbols we can write

$$(A)\frac{d(Z)}{d(A)} = \text{numerator of Eq. (10.4)}. \tag{10.5}$$

This relationship can be used to advantage to derive some alternative forms of Eq. (10.4) of general or particular applicability.

b. Equivalent algebraic transforms. The fundamental theorem of algebra states that a polynomial of degree n has n (and only n) roots. Specifically for the polynomial Z, the fundamental theorem of algebra may be stated as follows,

$$Z = 1 + K_1 A + K_1 K_2 A^2 + \cdots + \left(\prod_{l=1}^{n} K_l\right) A^n \tag{10.6}$$

$$= \left(\prod K_l\right)[(A - a_1)(A - a_2) \cdots (A - a_n)], \tag{10.7}$$

where $a_1, a_2, \ldots, a_n$ are the roots of Eq. (10.6). Inserting Eq. (10.5) into (10.4) we obtain

$$B = \frac{A\,(dZ/dA)}{Z} = A\frac{d\ln Z}{dA}, \tag{10.8}$$

which applied to Eq. (10.7) gives

$$B = A\left[\frac{d\ln(\prod_1^n K_l)}{dA} + \frac{d\ln(A - a_1)}{dA} + \frac{d\ln(A - a_2)}{dA} + \cdots\right]$$

$$= \frac{A}{A - a_1} + \frac{A}{A - a_2} + \cdots + \frac{A}{A - a_n} \tag{10.9}$$

or

$$B = \frac{(-1/a_1)A}{1 + (-1/a_1)A} + \frac{(-1/a_2)A}{1 + (-1/a_2)A} + \cdots + \frac{(-1/a_n)A}{1 + (-1/a_n)A}. \tag{10.10}$$

The roots $a_1, a_2, \ldots, a_n$ are constants for a specific polynomial Z, and hence we shall replace each $(-1/a_i)$ in Eq. (10.10) by a constant K_α, $K_\beta, \ldots, K_\nu$:

$$B = \frac{K_\alpha A}{1 + K_\alpha A} + \frac{K_\beta A}{1 + K_\beta A} + \cdots + \frac{K_\nu A}{1 + K_\nu A}. \tag{10.11}$$

This equation is rigorously equivalent to Eq. (10.4) as a representation for

A. Multiple Equilibria

B, despite its difference in appearance and despite the fact that the K_α, $K_\beta, \ldots, K_\nu$ do *not* correspond to $K_1, K_2, \ldots, K_n$, respectively.

c. Ideal circumstances: successive stoichiometric steps have same intrinsic affinity.

We can envision a particularly simple situation in which each site on the receptor is constituted of exactly the same atoms in the same structural configuration and that each such locus is completely unaffected by ligand binding at any other site. In other words each site acts as an isolated independent species, identical with each other site of the ensemble.

Under these circumstances, the stoichiometric K_i values, although *not* identical, are related to each other by some simple statistical factors. These we can generate by the following argument.

Let us define an *intrinsic* stoichiometric equilibrium constant K as one associated with a completely separated isolated monovalent binding site:

$$\text{Monovalent:} \quad -P + A \rightleftharpoons PA; \quad K = \frac{(PA)}{(P)(A)}. \tag{10.12}$$

We may represent a divalent binder by two binding steps each with an associated equilibrium constant:

$$\text{Divalent:} \quad \begin{aligned} -P- + A &\rightleftharpoons PA_1; & K_1, & \quad (10.13)\\ PA_1 + A &\rightleftharpoons PA_2; & K_2. & \quad (10.14)\end{aligned}$$

Any equilibrium constant, K_{equil} is also related* to the forward and reverse rate constants of the chemical transformation:

$$K_{\text{equil}} = k_{\text{forward}}/k_{\text{reverse}}. \tag{10.15}$$

If the binding affinity of P for A is intrinsically the same in Eqs. (10.13) and (10.12), the actual number of effective collisions between P and A leading to PA in the former (with two open sites) will be twice as great as in the latter (with only one open site) and hence so will the forward rate. On the other hand, for the reverse reaction, the thermal vibrations of the P–A bond and, hence, the rate of dissociation of PA into P and A will be the

---

* The validity of Eq. (10.15) can be demonstrated as follows. Consider the forward and reverse reactions of Eq. (10.12). The rate of the former is

$$\text{forward rate} = k_{\text{forward}} \, (P)(A)$$

and the latter rate is

$$\text{reverse rate} = k_{\text{reverse}} \, (PA).$$

At equilibrium

$$\text{forward rate} = \text{reverse rate}.$$

Equation (10.15) follows directly.

same in Eq. (10.13) as in Eq. (10.12) since each ligand–receptor bond is independent and identical. Thus, we conclude that

$$K_1 = \tfrac{2}{1} K. \tag{10.16}$$

By a similar argument we are led to the relation

$$K_2 = \tfrac{1}{2} K. \tag{10.17}$$

Turning to the general, n-valent association of P with A,

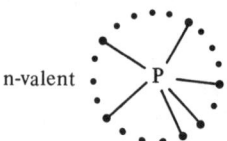

we recognize that the first mole of A to be bound to form PA has n times as many effective collisions in the forward rate as does A in Eq. (10.12). On the other hand, once the PA species is formed, the reverse rate depends on the vibrational dissociation of a single bond, independent of and identical with any other ligand–receptor bond. Hence,

$$K_1 = (n/1) K. \tag{10.18}$$

Once PA is formed, one mole of sites is occupied and $(n-1)$ moles are still open and unoccupied. Thus when $PA_1 + A \rightarrow PA_2$, the incoming A has $(n-1)$ times as many effective collisions as it does in the forward rate of Eq. (10.12). Once PA_2 exists, two ligand-receptor bonds can be broken by vibrational motions to regenerate PA_1, hence there is twice as great a likelihood of dissociation as compared to that in Eq. (10.12). We conclude, therefore, that

$$K_2 = [(n-1)/2] K \tag{10.19}$$

Continuing this type of analysis to the binding of the ith mole of A, we find

$$K_i = [n - (i-1)/i] K. \tag{10.20}$$

At the final saturation step when the nth mole of A is bound

$$K_n = (1/n) K \tag{10.21}$$

Equations (10.18)–(10.21) thus describe the statistical relations between successive binding constants when each mole of ligand is bound with identical intrinsic affinity (measured by K) because the sites are identical and independent.

Under these ideal circumstances, the stoichiometric binding equation [Eq. (10.4)] can be drastically simplified. For K_1, K_2, etc. in Eq. (10.4) we insert the appropriate statistical relationships obtainable from

A. Multiple Equilibria

Eqs. (10.18)–(10.20). Thus, we arrive at

$$B = \frac{\frac{n}{1}(KA) + 2\frac{n}{1}\frac{n-1}{2}(KA)^2 + \cdots + i\frac{n(n-1)(n-2)\cdots(n-i+1)}{1\cdot 2\cdot 3\cdots i}(KA)^i + \cdots}{1 + n(KA) + \frac{n(n-1)}{2!}(KA)^2 + \cdots + \frac{n(n-1)\cdots(n-i+1)(KA)^i}{i!} + \cdots + \frac{n(n-1)(n-2)\cdots(1)(KA)^n}{n!}} \quad (10.22)$$

As some may recognize, the denominator Z of Eq. (10.22) is the binominal expansion of

$$Z = [1 + (KA)]^n. \quad (10.23)$$

To find the numerator under these circumstances we turn back to Eq. (10.5):

$$\text{numerator} = A\{n[1 + (KA)]^{n-1}K\}. \quad (10.24)$$

Thus, the equation for moles bound can be reduced to the very simple form:

$$B = nKA/(1 + KA). \quad (10.25)$$

Note that for a monovalent system ($n = 1$)

$$B = KA/(1 + KA) \quad (10.26)$$

Thus, the n-valent ensemble behaves like n separate, independent, monovalent binding entities.

2. Site Formulation

Now let us recognize explicitly that receptors have molecular structures, in many cases known in refined detail, so that at least in principle we can designate specific sites as the loci of binding of ligand molecules. This view opens an alternative mode of analysis of the ligand–receptor equilibria. Such an analysis does not lend itself to as broad and general an algebraic formulation as the stoichiometric one, so we must specify clearly the constraints under which we can apply the equations derived.

a. Sites have different but fixed affinities. Recognizing the molecular nature of the receptor, we designate each site by an index number*,

---

* The site number is indicated by the index j placed as a left-hand subscript on each P. With this notation we can avoid confusion with the number of moles of bound ligand, i, of Eq. (10.1). In Eq. (10.27) each site can bind a maximum of 1A.

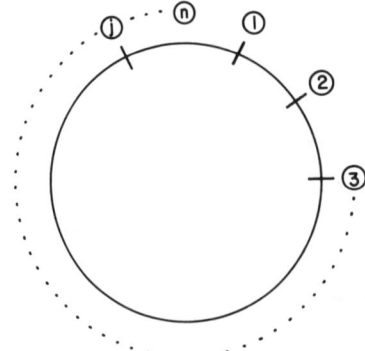

Fig. 10.1. Sites on a receptor indexed by numerical designation $1, 2, \ldots, j, \ldots, n$.

$1, 2, \ldots, j, \ldots, n$ (Fig. 10.1). For each site $_jP$ we can write a site equilibrium constant k_j:

$$
\begin{aligned}
_1P + A &\rightleftharpoons {_1PA}, & k_1 &= \frac{(_1PA)}{(_1P)(A)}; \\
_2P + A &\rightleftharpoons {_2PA}, & k_2 &= \frac{(_2PA)}{(_2P)(A)}; \\
&\vdots & &\vdots \\
_jP + A &\rightleftharpoons {_jPA}, & k_j &= \frac{(_jPA)}{(_jP)(A)}; \\
&\vdots & &\vdots \\
_nP + A &\rightleftharpoons {_nPA}, & k_n &= \frac{(_nPA)}{(_nP)(A)}.
\end{aligned} \qquad (10.27)
$$

For each site (e.g., number 1) $_1P$ represents all species of protein in which site 1 is empty (not occupied by ligand A). Some $_1P$'s may have site 2 occupied, others may have site 2 open; both states of site 2, site 3, etc. are encompassed in $_1P$. Corresponding statements apply to site-occupied form $_1PA$ which encompasses all species in which ligand is bound to site 1 regardless of the state of occupancy of the other sites.

For any single site (e.g., number 1) the moles of bound A (B_1) may be expressed as follows:

$$ B_1 = \frac{\text{moles A bound at site 1}}{\text{moles total protein}} = \frac{(_1PA)}{(_1P) + (_1PA)}. \qquad (10.28) $$

From the first site equilibrium constant in Eq. (10.27), we can obtain the relation $(_1PA) = k_1(_1P)(A)$ which on substitution into Eq. (10.28) leads to

$$ B_1 = \frac{k_1(_1P)(A)}{(_1P) + k_1(_1P)(A)} = \frac{k_1 A}{1 + k_1 A}. \qquad (10.29) $$

B. Graphical Representations

Corresponding statements can be written for site $2, 3, \ldots, j, \ldots, n$ so that, in general, we can say

$$B_j = \frac{k_j A}{1 + k_j A}. \tag{10.30}$$

If we add the moles of bound ligand at site 1 to that bound at site 2, site 3, etc., we have the total moles of bound ligand B.

$$B = \sum_1^n B_j = \frac{k_1 A}{1 + k_1 A} + \frac{k_2 A}{1 + k_2 A} + \cdots + \frac{k_j A}{1 + k_j A} + \cdots + \frac{k_n A}{1 + k_n A}. \tag{10.31}$$

Superficially, Eq. (10.31) resembles the algebraic transform equation [Eq. (10.11)], but it is *not* the same, because, in general, $K_\alpha \neq k_1, \ldots, K_\nu \neq k_n$.

b. Ideal circumstances: identical sites. Once again we consider an idealized limiting case, all sites identical in nature and with affinities that are fixed, i.e., nonchanging with occupancy of sites by ligands. Under these circumstances, in Eq. (10.31), $k_1 = k_2 = \cdots = k_j = \cdots = k_n$. Obviously, then Eq. (10.31) can be reduced to

$$B = nkA / (1 + kA). \tag{10.32}$$

Comparison of this result with that in Eq. (10.26) shows that

$$k = K,$$

that is, when the binding sites are identical (and nonchanging in affinity), the site equilibrium constant and the stoichiometric intrinsic equilibrium constant are identical.

We now have the mathematical frameworks for correlating ligand–receptor binding data. Let us turn then to actual representations and treatment of experimental data.

B. GRAPHICAL REPRESENTATIONS

To understand the biological action of an effector, we need to know the amount bound as a function of the free effector concentration, A, or F;[*] i.e., $B = g(F)$. If the necessary binding data have been collected, then $B = g(F)$ can be presented graphically or analytically in terms of an algebraic equation. We examine some graphical presentations first.

---

[*] The symbol F is widely used in the biological literature for the concentration of free ligand A, and we will use either symbol.

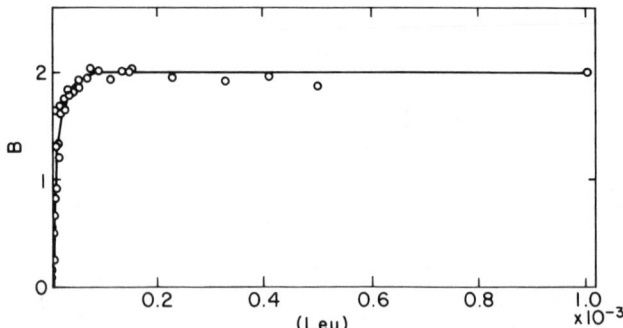

Fig. 10.2. Graphical representation of data for binding of leucine by α-isopropylmalate synthase using linear scales for experimental variables B and A. [From E. Teng-Leary and G. B. Kohlhaw, *Biochemistry* **12**, 2980 (1973).]

1. General Forms

Perhaps the most direct presentation is a plot of B versus F, as shown in Fig. 10.2 for the binding of leucine by the enzyme α-isopropylmalate synthase. When a wide range of experimental observations has been assembled, as in Fig. 10.2, many of the points are compressed into a narrow region and the variation of B with F is not readily discernible.

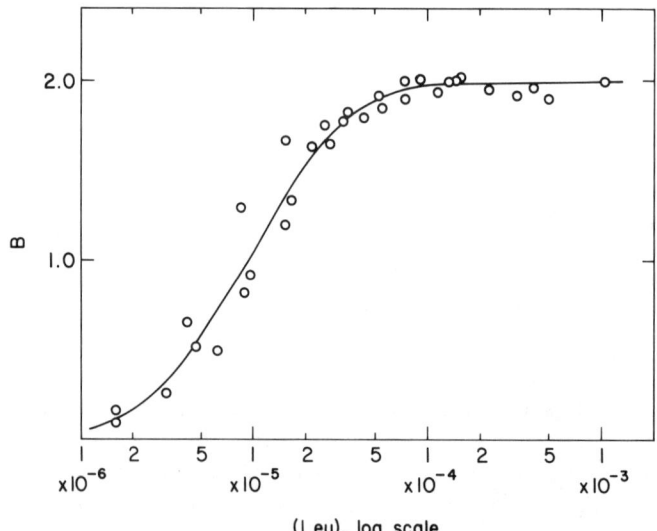

Fig. 10.3. Semilogarithmic representation of data for binding of leucine by α-isopropylmalate synthase.

B. Graphical Representations

Consequently, a graph of B versus $\log F$ (Fig. 10.3), is preferable because the experimental data are then spread out uniformly. This type of plot is analogous to an acid (or base) titration curve, where again binding (but of H^+ ion) is measured over a broad expanse of free proton concentrations. We can readily follow the course of B as free effector increases in concentration.

2. Ideal Case

Under ideal circumstances, when binding sites of a receptor are identical and nonchanging in affinity as occupancy increases, the semi-logarithmic plot acquires some special features (Fig. 10.4): (1) it is symmetrical about an inflection point (+ in Fig. 10.4); (2) the value of B at the inflection point is half of the saturation value n; (3) the value of $\log (A_{n/2})$ at the inflection point is $-\log K$ or $-\log k$.

Furthermore, under ideal circumstances it is possible to transform the ideal binding equation [Eq. (10.26) or (10.32)] into linear forms. For example, if we invert both sides of Eq. (10.25) we obtain

$$\frac{1}{B} = \frac{1}{nK}\frac{1}{A} + \frac{1}{n}. \tag{10.33}$$

Obviously a graph of $1/B$ versus $1/A$ (Fig. 10.5a) is linear with an intercept on the ordinate axis of $1/n$ and a slope of $1/nK$. Alternatively, if both sides of Eq. (10.33) are multiplied by A, we obtain

$$\frac{A}{B} = \frac{1}{nK} + \frac{1}{n}A. \tag{10.34}$$

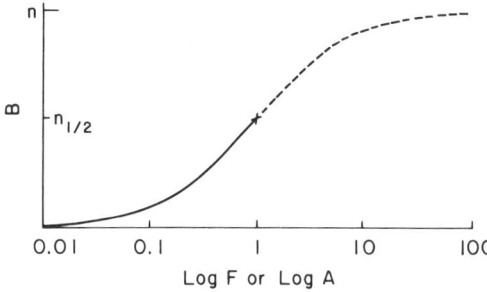

Fig. 10.4. Semilogarithmic graph (B vs. log F) for a receptor with *n identical binding sites* for ligand. For this ideal system, the amount bound at the inflection point (+) will be one-half that at saturation (upper plateau). The curve is shown as a broken line in the region where data are lacking in most published binding studies.

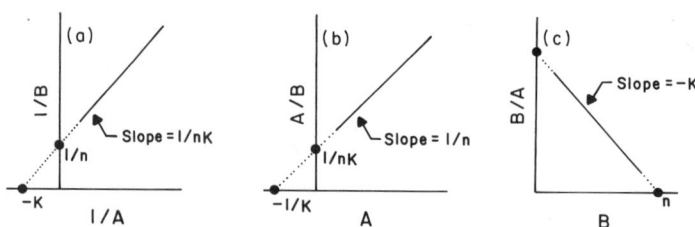

Fig. 10.5. Graphical representations of the linear transforms of the ideal binding equation for a receptor with n identical binding sites. The linear equations giving the appropriate coordinates for these graphs [(a), $1/B$ versus $1/A$; (b) A/B versus A; and (c) B/A versus B] were all derived by Wolff in 1929 (see, for example, J. B. S. Haldane and K. G. Stern, "Allgemeine Chemie de Enzyme," p. 119. Steinkopf, Dresden, 1932.) In the succeeding decades many other proper names have been attached to these graphs.

Thus a graph of A/B versus A (Fig. 10.5b) is linear with an intercept on the ordinate axis of $1/nK$ and a slope of $1/n$. Still another transform is obtained if we move the denominator from the right side of Eq. (10.25) to the left side so that

$$B(1 + KA) = nKA, \tag{10.35}$$

which can then be rearranged to

$$B = nKA - BKA. \tag{10.36}$$

This in turn gives

$$B/A = nK - KB, \tag{10.37}$$

which corresponds to a linear graph (Fig. 10.5c) when the coordinates are B/A and B; in this graph the intercept on the abscissa is n and that on the ordinate nK.

All three linear transforms were described by Wolff over fifty years ago. They have been used in various experimental contexts and have accumulated a number of names as designations: Langmuir and Lineweaver–Burk for Fig. 10.5a; Hanes, Scott, and Hildebrand–Benesi for Fig. 10.5b; Eadie, Eadie–Hofstee, and Scatchard for Fig. 10.5c.

Figure 10.6 shows some published data for ligand binding by benzodiazepine receptors. A straight line has been drawn through the points in Fig. 10.6a, and n, the intercept on the abscissa, is estimated as slightly over 0.9. This graph is concordant with Fig. 10.6b which has the symmetrical shape characteristic of a single set of identical sites and an inflection point near $B = 0.5$ and which approaches a plateau for B at slightly above 0.9 as F increases. The binding constants n and K can thus be extracted from Fig. 10.6a or from Fig. 10.6b.

B. Graphical Representations

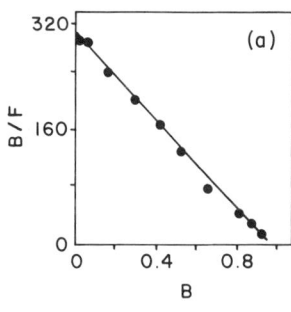
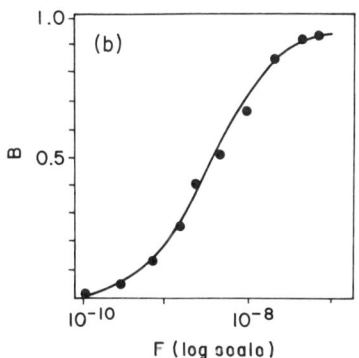

Fig. 10.6. Binding of a benzodiazepine by membrane receptors. [Adapted from A. C. Bowling and R. J. De Lorenzo, *Science* **216**, 1247 (1982).]

It should be recognized, however, that when data are plotted on a B/F versus B graph there is an enormous temptation to fit them to a straight line, either by eye or by least squares methods. In most cases it can be shown readily that the conclusions derived from the B/F versus B graph are untenable. For example, for a different system of benzodiazepine receptors, binding of a ligand drug has been found to give the graph of Fig. 10.7a. Individual points have been fitted by a straight line with a very good correlation coefficient, and the total number of receptor sites (830) computed from the intercept on the abscissa. The last experimental point (Fig. 10.7a), at $B \sim 700$ is situated, presumably, about 15% before the saturation value. Let us examine the same data as points on a graph of B versus the logarithm of free drug (Fig. 10.7b). Since the graph of Fig. 10.7a has been purported to be a straight line, the semilogarithmic plot (Fig. 10.7b) should be an ideal S-shaped curve. Clearly, actual experimental data are available for less than half of the S; the curve has not reached an inflection point, and obviously there is not even a hint of the slope decreasing to approach a plateau value of n. The experimental point at $B \sim 700$ cannot possibly be within 15% of the saturation value. If we arbitrarily assume that the highest observed value of B is at the inflection point, then n must be at least double this value, or 1400, and is likely to be even higher. Clearly the value of 830 deduced from the B/F versus B plot is nowhere near the actual number of receptor sites.

3. Deviations from Ideal, Statistical Behavior

It is convenient to visualize nonideal binding in terms of deviations from the ideal curve. Let us start with the semilogarithmic form (Fig. 10.8). If the actual binding curve falls uniformly above the ideal, the affinity must be

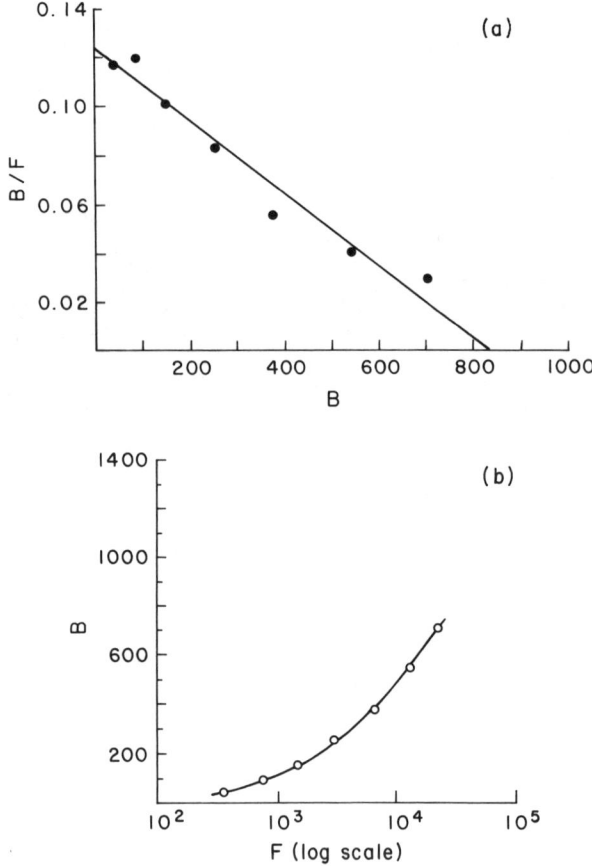

Fig. 10.7. Binding of diazepam by a different system of receptors. [Adapted from P. Skolnick, V. Moncada, J. L. Barker, and S. M. Paul, *Science* **211**, 1448 (1981).]

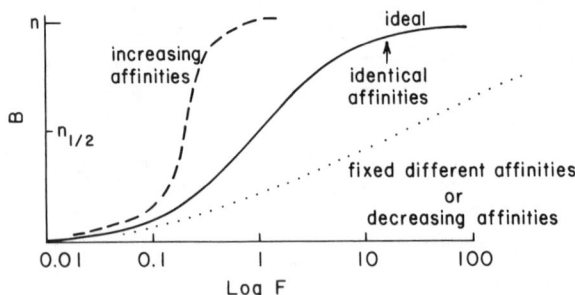

Fig. 10.8. Semilogarithmic graph of ligand–receptor binding in nonideal cases, where actual binding is greater or less than that for identical binding sites.

B. Graphical Representations

increasing (above the statistical value) for each step. We speak then of "cooperative" behavior and picture variations in conformation of the receptor that accentuate binding affinities as occupancy by the ligand increases.

Alternatively, the actual binding curve may fall below the ideal (Fig. 10.8). Such behavior is a manifestation of either of the following situations: (1) The affinities may be fixed but different from statistical, that is, there may be different classes of sites with fixed affinities; (2) the sites may originally be all identical but at least some affinity decreases with increasing occupancy by ligand. Not all sites need decrease in affinity; in fact some stages in binding may even show increases with increasing occupancy so long as others decrease to more than a compensating degree. The behavior described by (2) is sometimes called "negative cooperativity." It should be noted that from binding data alone, (2) cannot be distinguished from (1), that is, negative cooperativity cannot be differentiated from a collection of heterogeneous sites of fixed affinities.

As we have seen (Fig. 10.5), in a graph of B/F versus B, ideal behavior manifests itself as a straight line. In principle, accentuated affinities, stronger than ideal, should lead to a curve lying above the ideal line (Fig. 10.9), and attenuated affinities, less than ideal, should generate a curve below the line.

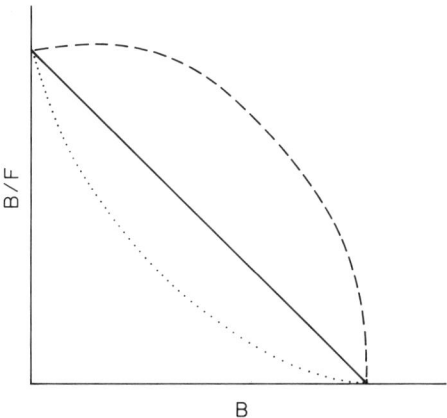

Fig. 10.9. Course of Scatchard graph for receptor behaving ideally with all n binding sites identical and unchanging in affinity(———); receptor for which binding affinities increase with increasing occupancy of sites (– – –); and receptor for which either sites have different but fixed affinities, unchanging with occupancy, or sites are originally identical (or different) but their affinities for ligand decrease with increasing occupancy of sites (· · · ·).

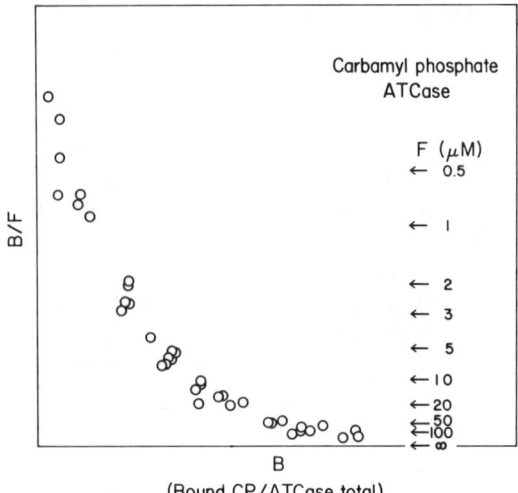

Fig. 10.10. Scatchard graph for binding of carbamoyl phosphate by aspartate transcarbamoylase. To emphasize the increasingly severe compression of the ordinate axis, B/F at left, corresponding values of F are shown at right.

The latter concave curve can entice us toward an intercept on the B axis that may be seriously unrealistic. An inherent problem with the B/F versus B graph is that wide ranges of experimental data are compressed into a small region of the B/F coordinate. For example, as Fig. 10.10 shows, for values of F from 10^{-7} to 10^{-5} M, we cover more than 80% of the total span of data on the ordinate axis but for values of F from 10^{-4} to 10^{-1} M less than 5%. If we happen to know the number of binding sites from extra-thermodynamic structural investigations, it is easy to convince ourselves that binding data point to the appropriate intercept on the B axis. For this reason, in Fig. 10.10, the numbers along the abscissa have been blotted out; an objective observer notices immediately that the data approach the x axis with zero slope. In general, therefore, a concave curve on a B/F versus B graph should not be used to estimate saturation number.

C. ANALYTICAL REPRESENTATIONS OF DATA

In essence what we need is to evaluate the stoichiometric equilibrium constants of Eq. (10.4), the simulator equilibrium constants of Eq. (10.11), or, when possible, the site constants of Eq. (10.31). In

C. Analytical Representations of Data

general, such evaluations can be made by appropriate statistical fitting and allied computer programs.*

1. Evaluation of K_1

It is almost always possible to obtain K_1 by a simple graphical procedure. As is evident from Eq. (10.4), after we factor out (A) from the numerator and transfer it to the left-hand side of the equation,

$$\lim_{(A) \to 0} \frac{B}{(A)} = \lim \frac{B}{F} = K_1. \tag{10.38}$$

In general, it is possible to make measurements with good precision at low concentrations of ligand. Hence an extrapolation such as is illustrated for a specific example in Fig. 10.11 provides a value for the first stoichiometric equilibrium constant.

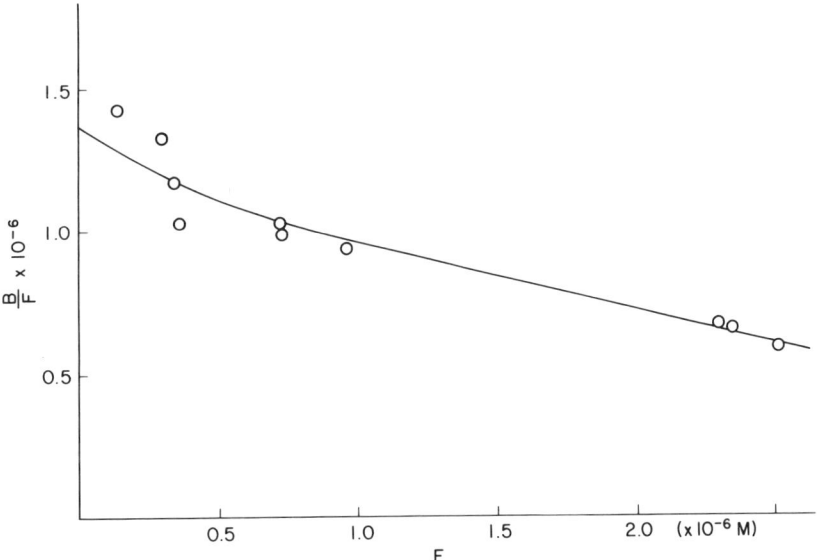

Fig. 10.11. Evaluation of first stoichiometric binding constant (for binding of carbamoyl phosphate by aspartate transcarbamoylase) by extrapolation of B/F to limiting value as free ligand concentration approaches zero [see Eq. (10.4)].

* See, for example, J. E. Fletcher, A. A. Spector, and J. D. Ashbrook *Biochemistry* **9**, 4580–4587 (1970); H. A. Feldman *Anal. Biochem.* **48**, 317–338 (1972); P. J. Munson and D. Rodbard *Anal. Biochem.* **107**, 220–239 (1980); B. Honoré and R. Brodersen, *Mol. Pharmacol.* **25**, 137–150 (1984).

2. Relationships Between Stoichiometric and Site Binding Constants When Sites Have Fixed But Different Affinities

When the sites have fixed affinities, any one of the Eqs. (10.4), (10.11), or (10.31) can be used to correlate the experimental data. The different types of equilibrium constants, therefore, must be related to each other, and any one type can be determined from the other. Let us first derive the relationships between equilibrium constants for the simplest, two-site system and then generalize to multisite receptors.

a. Two-site system. A schematic diagram of a two-site system (Fig. 10.12) helps in visualization of interrelationships between constants. The equilibrium pathways to ligand-receptor species $_1$PA, $_2$PA, and $_{1,2}$PA$_2$, in which sites 1, 2, and both, respectively, are occupied by ligand, are shown with the corresponding site equilibrium constants. Below the site equilibria are schematic equations showing the species involved in the stoichiometric equilibria.

Since conservation of mass requires that

$$(PA_1) = (_1PA) + (_2PA), \tag{10.39}$$

we can write

$$K_1 = \frac{(PA_1)}{(P)(A)} = \frac{[(_1PA) + (_2PA)]}{(P)(A)} = k_1 + k_2. \tag{10.40}$$

From the set of relations assembled in Eq. (10.1) we find that

$$K_1 K_2 = \frac{(PA_2)}{(P)(A)^2}, \tag{10.41}$$

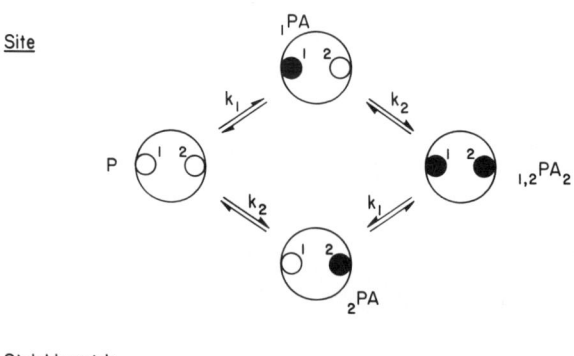

Fig. 10.12. Comparison of meanings of stoichiometric and site binding constants for a two-site receptor in which sites have different but fixed affinities.

C. Analytical Representations of Data

which can be manipulated into

$$K_1 K_2 = \frac{(PA_2)}{(P)(A)^2} \frac{(_1 PA)}{(_1 PA)} = \frac{(PA_2)}{(_1 PA)(A)} \frac{(_1 PA)}{(P)(A)} = k_1 k_2. \tag{10.42}$$

Thus, we have two equations relating K_1 and K_2 to k_1 and k_2, so if either pair has been evaluated from experimental data the other pair can be promptly calculated.

If we write Eqs. (10.4) and (10.11) for a divalent system,

$$\frac{K_1 A + 2 K_1 K_2 A^2}{1 + K_1 A + K_1 K_2 A^2} = B = \frac{K_\alpha A}{1 + K_\alpha A} + \frac{K_\beta A}{1 + K_\beta A}, \tag{10.43}$$

then with some modest algebraic manipulation* we can obtain

$$K_1 = K_\alpha + K_\beta, \tag{10.44}$$

$$K_1 K_2 = K_\alpha K_\beta. \tag{10.45}$$

These relations are always valid for a divalent ligand–receptor system, regardless of interactions between sites. Nevertheless, when sites have fixed affinities and, hence, site constants that are specified by Eqs. (10.40) and (10.42), then it is apparent from a comparison with Eqs. (10.44) and (10.45) that

$$K_\alpha = k_1, \tag{10.46}$$

$$K_\beta = k_2. \tag{10.47}$$

In other words, under these circumstances the ghost sites α and β are reified, that is, are real sites on the receptor. Under these circumstances, there is no need for separate binding constants K_α and K_β.

b. Three-site system. By algebraic procedures analogous to those used with a divalent system,* we can derive the following relations between equilibrium constants:

$$K_1 = k_1 + k_2 + k_3, \tag{10.48}$$

$$K_1 K_2 = k_1 k_2 + k_1 k_3 + k_2 k_3, \tag{10.49}$$

$$K_1 K_2 K_3 = k_1 k_2 k_3, \tag{10.50}$$

Again, if either triplet of constants has been evaluated from experimental data, the other group can be promptly calculated.

---

* See, for example, I. M. Klotz and D. L. Hunston, *Arch. Biochem. Biophys.* **193**, 314–328 (1979); I. M. Klotz and D. L. Hunston, *J. Biol. Chem.* **250**, 3001–3009 (1975).

c. Multisite system. By visual analogy with Eqs. (10.40)–(10.42) and (10.48)–(10.50) for two- and three-site systems, respectively, or preferably by algebraic operations,* we can obtain appropriate expressions relating stoichiometric and site constants for the general case of a multisite receptor:

$$K_1 = k_1 + k_2 + \cdots + k_i + \cdots + k_n, \tag{10.51}$$

$$K_1 K_2 = k_1 k_2 + k_1 k_3 + \cdots + k_1 k_n + k_2 k_3 + k_2 k_4 + \cdots + k_{n-1} k_n,$$
$$\vdots \tag{10.52}$$

$$K_1 K_2 \cdots K_n = k_1 k_2 \cdots k_n. \tag{10.53}$$

Again, if either group (K_i or k_j) has been evaluated, the other can be computed, for we have n equations for n unknowns.

3. Binding Constants When Site Affinities Change With Extent of Occupancy

Now we must be more careful in indexing site binding constants, for affinity at a given site changes when other sites are occupied by ligand. To facilitate our analysis, let us start with a divalent ligand–receptor system.

a. Two-site system. In Fig. 10.13 we[†] see that there are two alternative avenues for proceding from ligand-free receptor P to fully saturated receptor $_{1,2}PA_2$. The first ligand could occupy site 1 to give $_1PA$; the corresponding site equilibrium constant is k_1. A second ligand can then occupy site 2 of $_1PA$ to give $_{1,2}PA_2$. For this second step the site equilibrium constant is not k_2, for the latter is a measure of the affinity of site 2 in P, i.e., when site 1 is empty. However, in $_1PA$ site 1 is occupied and can exert an effect on the affinity for ligand of site 2. Therefore, the site equilibrium constant must be designated by an index $k_{1,2}$ that clear distinguishes if from k_2. For the alternative pathway from empty P to saturated $_{1,2}PA_2$ we define two site equilibrium constants k_2 and $k_{2,1}$ (Fig. 10.13).

Insofar as the stoichiometric equilibria are concerned, however, we need not take cognizance of the sites. For uptake of the first mole of ligand, we can define K_1, for the second mole, K_2 (Fig. 10.13). From general

---

* See, for example, I. M. Klotz and D. L. Hunston, *Arch. Biochem. Biophys.*, **193**, 314–328 (1979); I. M. Klotz and D. L. Hunston, *J. Biol. Chem.* **250**, 3001–3009 (1975).

[†] The figure shows the species involved in site, stoichiometric and simulator, "ghost," equilibria. Since each of the equilibrium constants K_α, K_β, etc. does not refer to any specific actual site on the receptor, we can assign each to a "ghost" site. The ensemble of n imaginary, nonexistent ghost sites with corresponding simulator equilibrium constants faithfully reproduces the observed dependence of B on F.

C. Analytical Representations of Data

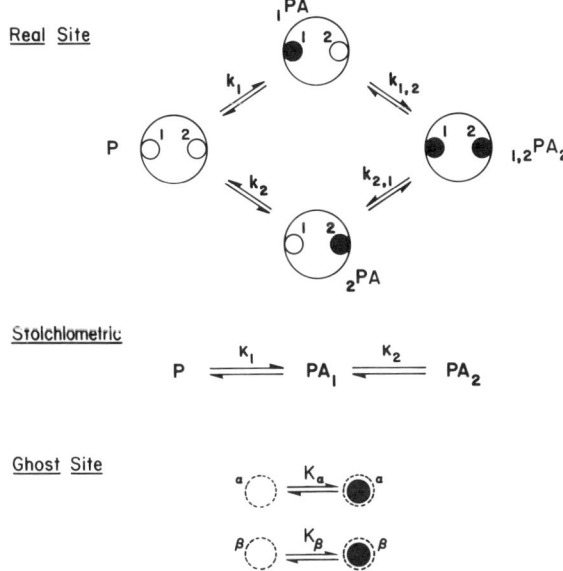

Fig. 10.13. Comparison of meanings of different types of ligand–receptor binding constants for a two-site receptor. Distinctions must be made between the stoichiometric formulation and the site formulations in terms of *real* or *ghost* sites.

algebraic arguments described earlier we also define two simulator equilibrium constants, K_α and K_β, for the corresponding ghost sites (Fig. 10.13).

Thus, it is evident that a bivalent ligand–receptor system has assigned to it four site constants but only two stoichiometric constants. Actually only three of the four site constants are independent. By inserting the appropriate molecular species into the product constants k_1k_{12} and k_2k_{21},

$$k_1k_{12} = \frac{(_1\text{PA})}{(\text{P})(\text{A})} \frac{(_{1,2}\text{PA}_2)}{(_1\text{PA})(\text{A})}; \quad k_2k_{21} = \frac{(_2\text{PA})}{(\text{P})(\text{A})} \frac{_{1,2}\text{PA}_2}{(_2\text{PA})(\text{A})}, \quad (10.54)$$

we can see that

$$k_1k_{1,2} = k_2k_{21}.$$

Thus, if any three of the site constants are known, the fourth can be calculated. Nevertheless, three out of the four are independent parameters.

We can still find relations between the stoichiometric equilibrium constants and the site ones. Thus, for the first stoichiometric step

$$K_1 = \frac{(\text{PA})}{(\text{P})(\text{A})} = \frac{[(_1\text{PA}) + (_2\text{PA})]}{(\text{P})(\text{A})} = k_1 + k_2, \quad (10.55)$$

and for the second

$$K_1K_2 = \frac{(PA_2)}{(P)(A)^2} = \frac{(PA_2)}{(P)(A)^2}\frac{(_1PA)}{(_1PA)} = \frac{(_1PA)}{(P)(A)}\frac{(PA_2)}{(_1PA)(A)} = k_1k_{1,2}. \quad (10.56)$$

However, now we are faced with an indeterminacy in the site constants. The stoichiometric constants K_1 and K_2 could be evaluated for a given set of binding data by fitting them to the stoichiometric binding equation [Eq. (10.43)]. However, there is no way of specifying all three site constants from the two equations [Eqs. (10.55) and (10.56)]. A third relation between site constants is supplied by Eq. (10.54) but simultaneously it introduces a fourth variable. Thus, from binding data alone there is no way of establishing values for the three independent site constants; they can exhibit "free will," i.e., one of them can assume any value so long as the other two are constrained by the relations of Eqs. (10.55) and (10.56).

When site affinities change with occupancy, then it is apparent that Eq. (10.31) cannot be correct. If we insert all the site constants of Fig. 10.13, we obtain

$$B = \frac{k_1 F}{1 + k_1 F} + \frac{k_2 F}{1 + k_2 F} + \frac{k_{1,2} F}{1 + k_{1,2} F} + \frac{k_{2,1} F}{1 + k_{2,1} F}. \quad (10.57)$$

However, each term on the right-hand side in Eq. (10.57) approaches 1 as $F \to \infty$; hence the four terms approach 4. However in a divalent system the maximum value of B is 2. Obviously a site-binding equation of the format of Eq. (10.57) is meaningless.

Nevertheless Eq. (10.11) [or the right-hand side of Eq. (10.43)] for ghost sites with simulator binding constants K_α and K_β is fully valid. However, K_α or K_β is *not* simply k_1 or k_2 or $k_{1,2}$ or $k_{2,1}$.

How are K_α and K_β related to the site constants? For a bivalent ligand–receptor system explicit expressions are readily obtained:*

$$K_\alpha = \tfrac{1}{2}(k_1 + k_2) \pm \tfrac{1}{2}[(k_1 + k_2)^2 - 4k_1k_{1,2}]^{1/2}, \quad (10.58)$$

$$K_\beta = \tfrac{1}{2}(k_1 + k_2) \mp \tfrac{1}{2}[(k_1 + k_2)^2 - 4k_1k_{1,2}]^{1/2}. \quad (10.59)$$

Again we see that there are three site constants in these two equations, so all three *cannot* have specified values. On the other hand, if the stoichiometric constants have been evaluated from experimental binding data, then the two parameters K_α and K_β have explicit values* from the

---

* See, for example, I. M. Klotz and D. L. Hunston, *Arch. Biochem. Biophys.* **193**, 314–328 (1979); I. M. Klotz and D. L. Hunston, *J. Biol. Chem.* **250**, 3001–3009 (1975).

C. Analytical Representations of Data

pair of equations:

$$K_\alpha = \tfrac{1}{2}K_1 \pm \tfrac{1}{2}(K_1^2 - 4K_1K_2)^{1/2}, \tag{10.60}$$

$$K_\beta = \tfrac{1}{2}K_1 \mp \tfrac{1}{2}(K_1^2 - 4K_1K_2)^{1/2}. \tag{10.61}$$

As a concrete illustration of relationships between these binding constants, let us examine some results for the binding of leucine by isopropylmalate synthase. From experimental binding data, the stoichiometric binding constants have been found to be

$$K_1 = 0.48 \times 10^5, \qquad K_2 = 2.5 \times 10^5.$$

It is also perfectly feasible to determine the simulator binding constants from Eqs. (10.60) and (10.61):

$$K_\alpha = 2.4 \times 10^4 + 1.1 \times 10^5 \sqrt{-1}, \qquad K_\beta = 2.4 \times 10^4 - 1.1 \times 10^5 \sqrt{-1}$$

Thus, in this system the simulator binding constants are imaginary or complex numbers. Clearly they cannot be meaningful values for any actual site-binding constants. They are eminently appropriate, however, for ghost sites.

Can we say anything about the site-binding constants? If we propose some form of restraining molecular model then the answer may be yes. For example, the receptor isopropylmalate synthase is constituted of two identical protomers, as is known from experimental information derived from other than binding studies. Since the protomers are identical when no ligand is attached to the receptor, we can state that

$$k_1 = k_2 \equiv k_\mathrm{I}.$$

Furthermore (see Fig. 10.13) we would expect that

$$k_{1,2} = k_{2,1} \equiv k_\mathrm{II}.$$

Under these circumstances, we can use Eqs. (10.58) and (10.59) to show that

$$k_1 = 0.24 \times 10^5 = k_2, \qquad k_{1,2} = 5.0 \times 10^5 = k_{2,1}.$$

We might notice also that even though in our model *both binding sites are identical* at the outset, their binding of ligand fits a two-term equation [Eq. (10.43)] that *superficially looks like* an expression for two different classes of sites.

b. Multisite systems. If we turn to a trivalent ligand–receptor system, the number of site constants increases to 12 (Fig. 10.14), of which seven are independent. As Table 10.1 illustrates, the rise in the number of site constants is extremely rapid for further increases in number of moles

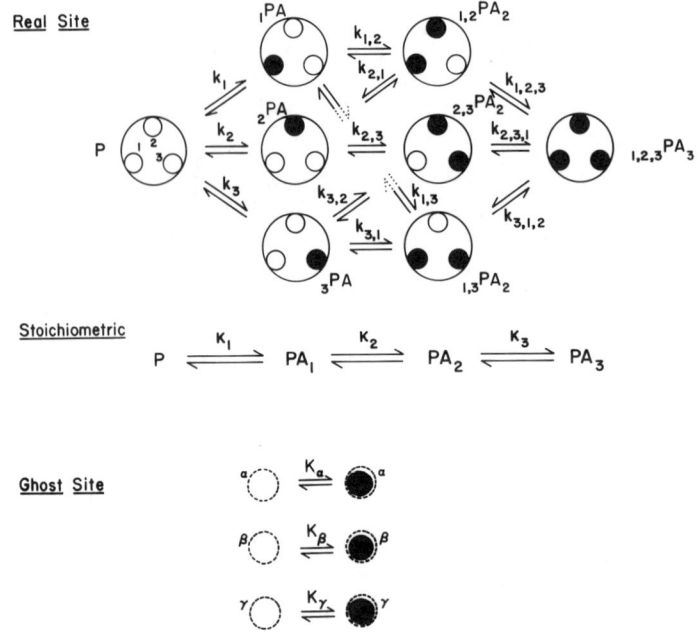

Fig. 10.14. Comparison of meanings of different types of binding constants for a three-site receptor.

TABLE 10.1

Comparison of Numbers of Different Kinds of Binding Constants for a Macromolecule with n Binding Sites

Number of binding sites, n	Total number of site binding constants	Number of independent site binding constants	Number of stoichiometric binding constants	Number of simulator binding constants
2	4	3	2	2
3	12	7	3	3
4	32	15	4	4
6	192	63	6	6
8	1,024	255	8	8
12	24,576	4,095	12	12

C. Analytical Representations of Data

of ligand bound per mole receptor system. Nevertheless, the number of stoichiometric (or of simulator) equilibrium constants is always equal to the stoichiometry of bound ligand. Thus, it would clearly be futile to attempt to establish values for the site constants, even after the stoichiometric constants had been evaluated from binding data, for there are many different combinations of assignments of k_j's that would fit the experimental observations. Even for a trivalent system, the seven independent k_j's cannot be fixed by three known experimental constants K_i.

On the other hand, if some molecular information is available, it may be possible to extract site constants from binding data once the stoichiometric constants have been evaluated. For example, the catalytic trimer of aspartate transcarbamoylase binds trifluorodihydroxypropyl phosphonate over a range of concentrations; from experimental observations the following stoichiometric constants have been computed:*

$$K_1 = 2.89 \times 10^4, \quad K_2 = 0.0295 \times 10^4, \quad K_3 = 0.0875 \times 10^4.$$

In the initially ligand-free trimer, all three monomer subunits are identical. Consequently we can state that

$$k_1 = k_2 = k_3 \equiv k_\mathrm{I}.$$

Once the first ligand has been bound, it affects the affinity of the open sites for the second molecule of ligand, that is, for the conversion of PA_1 to PA_2. If each occupied subunit has the same effect on the ligand affinities of the open subunits, we may write

$$k_{1,2} = k_{1,3} = k_{2,1} = k_{2,3} = k_{3,1} = k_{3,2} \equiv k_\mathrm{II}.$$

Finally, if the last open site has the same affinity for the third ligand regardless of which subunit it is on, but this affinity differs from that for uptake of the second ligand, then we can state that

$$k_{1,2,3} = k_{2,3,1} = k_{3,1,2} \equiv k_\mathrm{III}.$$

Now we have reduced the number of site parameters to three, and their numerical values can be established from the known constants K_1, K_2 and K_3:*

$$k_\mathrm{I} = 0.96 \times 10^4, \quad k_\mathrm{II} = 0.0295 \times 10^4, \quad k_\mathrm{III} = 0.262 \times 10^4.$$

It should be emphasized, however, that these values *are not site-binding constants* for sites 1, 2, and 3, respectively. No single value of a site-binding

---

* See, for example, I. M. Klotz and D. L. Hunston, *Arch. Biochem. Biophys.* **193**, 314–328 (1979); I. M. Klotz and D. L. Hunston, *J. Biol. Chem.* **250**, 3001–3009 (1975).

constant can be assigned to site 1 (or 2 or 3). The affinity at each site depends on whether the other sites are open or occupied.

For a tetravalent system there are 4 stoichiometric constants and 32 site constants, of which 15 are independent. Thus, there is much more latitude in the molecular scenarios that we can construct to place constraints on the system that might be in accord with features of the molecular structure of the receptor. One special case of interest is that in which initially identical subunits interact in a pairwise fashion, that is when one is occupied its unoccupied pair-partner changes in affinity but the unoccupied pair is unaffected. In this circumstance the site binding constants can be reduced* to just two parameters. The observed uptake of ligand may appear superficially to reflect the existence of two different classes of binding site* in the receptor even though all of the sites are initially identical.

There are many proteins for which the total number of binding sites cannot be determined, or may even be indeterminate. Nevertheless, ligand-receptor binding can be expressed analytically in terms of the stoichiometric constants. Binding by the serum protein albumin is especially illustrative of this situation. Many metabolites, hormones, drugs, and other ligands are carried by serum albumin, and the extent of binding of many of these substances has been measured quantitatively. For example, for diflunisal, an anti-inflammatory drug, the extensive binding by serum albumin can be correlated by fifteen stoichiometric constants:[†]

$K_1 = 500{,}000$, $K_4 = 75{,}000$, $K_7 = 16{,}000$, $K_{10} = 2700$, $K_{13} = 730$,

$K_2 = 220{,}000$, $K_5 = 47{,}000$, $K_8 = 7700$, $K_{11} = 1800$, $K_{14} = 420$,

$K_3 = 120{,}000$, $K_6 = 28{,}000$, $K_9 = 4400$, $K_{12} = 1200$, $K_{15} = 180$.

This ligand-receptor ensemble has 245,760 site binding constants.

D. THERMODYNAMIC QUANTITIES

Once equilibrium constants have been determined, it is a straightforward matter to calculate free energies for ligand-receptor binding (see Chapter 4). For example, if K_1 for the binding[†] of salicylate by serum albumin is 48,000 and that for diflunisal is 500,000, then at 37°C:

$$\Delta G_1^\circ = -RT \ln K_1 = -6650 \quad \text{cal/mole for salicylate}$$

$$= -8050 \quad \text{cal/mole for diflunisal.}$$

---

[*] See, for example, I. M. Klotz and D. L. Hunston, *Arch. Biochem. Biophys.* **193**, 314–328 (1979); I. M. Klotz and D. L. Hunston, *J. Biol. Chem.* **250**, 3001–3009 (1975).

[†] B. Honoré and R. Brodersen, *Mol. Pharmacol.* **25**, 137–150 (1984).

D. Thermodynamic Quantities

Thus, we can conclude that the additional substituents in a molecule of diflunisal as compared with salicylate, increase the interaction standard free energy of the first mole of ligand with serum albumin by -1400 cal/mole, and we may obtain some insight into molecular structural features that strengthen this (nonspecific) binding.

On the other hand, it is *not* meaningful to say that the interaction free energy has increased by 21% {[(8050–6650)/6650] × 100}. Such a claim implies that a $\Delta G°$ of zero corresponds to no interaction between ligand and receptor. That claim is clearly wrong for a $\Delta G° = 0$ corresponds to an equilibrium constant of 1 [see Eq. (4.1)]. Depending on the concentrations of ligand and receptor, a substantial fraction of receptor may be in the liganded state PA_1.

We can also compare free energies of successive stoichiometric binding steps. For example, for binding of leucine by isopropylmalate synthase

$$K_1 = 0.48 \times 10^5, \qquad K_2 = 2.5 \times 10^5.$$

Thus,

$$\Delta G_2° - \Delta G_1° = -RT \ln K_2 - (-RT \ln K_1)$$
$$= -980 \quad \text{cal/mole}.$$

Some of this free energy change reflects merely the statistics of the difference in number of open sites when one mole of ligand is already on the receptor as compared to receptor completely unliganded. For a divalent system, [see Eqs. (10.16) and (10.17)],

$$\left(\frac{K_2}{K_1}\right)_{\text{statistical}} = \frac{(1/2)K}{(2/1)K} = \frac{1}{4}.$$

Hence, it is convenient to separate the difference free energy into two terms:

$$\Delta(\Delta G°) = \Delta G_2° - \Delta G_1° = \Delta G_{\text{statistical}} + \Delta G_{\text{interaction}}. \tag{10.62}$$

In the present example with leucine–isopropylmalate synthase,

$$\Delta G°_{\text{statistical}} = -RT \ln \tfrac{1}{4} = 820 \quad \text{cal/mole}.$$

Thus, the corrected interaction free energy between successive binding steps is

$$\Delta G°_{\text{interaction}} = -980 - 820 = -1800 \quad \text{cal/mole}.$$

The negative $\Delta G°_{\text{interaction}}$ means that the binding of the first mole of ligand facilitates binding of the second mole. Such "cooperative" interactions are particularly well established in the uptake of oxygen by hemoglobins.

If ligand–receptor binding is measured over a range of temperature, we can calculate the $\Delta H°$ of complex formation from the thermodynamic relation*

$$\frac{d \ln K}{dT} = \frac{\Delta H°}{RT^2}.$$

Occasionally an alternative direct calorimetric study of the heat of binding has been made. In either event, once $\Delta H°$ has been established, the entropy change $\Delta S°$ can be computed from [see Eq. (3.4)]

$$\Delta G° = \Delta H° - T\Delta S°.$$

Using these methods, we find, for example, the following thermodynamic parameters for the binding (at 25°C) of the first mole of a small anionic dye, methyl orange, by serum albumin:

$\Delta G_1° = -6400$ cal/mole,

$\Delta H_1° = -2100$ cal/mole,

$\Delta S_1° = 14.5$ cal/mole deg.

Complexes of other small organic anions with serum albumin have similar properties. Hydrophobic interactions should be associated with a relatively small ΔH (compared to ΔG) and a large positive ΔS. Since the parameters for anion–albumin binding fit these characteristics, we are tempted to conclude that hydrophobic interactions are the molecular basis of the thermodynamic stability of the ligand–receptor complexes. Such a conclusion is not warranted, however. Electrostatic interactions are also accompanied by a small ΔH and a large positive ΔS. This can be shown from theoretical electrostatic considerations.[†] It is also obvious from many experimental investigations. For example, for the binding of Cu^{2+} ions by serum albumin (at 25°C)

$\Delta G_1° = -5900$ cal/mole,

$\Delta H_1° = 2800$ cal/mole,

$\Delta S_1° = +29$ cal/mole deg.

For metal ion–receptor binding it is difficult to invoke hydrophobic interactions.

---

  * I. M. Klotz and R. M. Rosenberg, "Chemical Thermodynamics," 3rd edition, p. 141. Benjamin, Menlo Park, California, 1972; 4th edition, 1986.
  † I. M. Klotz, *Ann. NY Acad. Sci.* **226**, 18–35 (1973).

In a complementary fashion, we can also cite examples of the binding of long apolar molecules by a receptor, where the thermodynamic data are concordant with the expectation that hydrophobic interactions should play a dominant role. Thus, for the binding of micelles of hexadecylphosphorylcholine by phospholipase A_2, the thermodynamic quantities reported are*

$$\Delta G° = -6800 \quad \text{cal/mole,}$$
$$\Delta H° = -360 \quad \text{cal/mole,}$$
$$\Delta S° = 24 \quad \text{cal/mole deg.}$$

Ultimately from all of these considerations we must turn to the central biological question: what is the physiological result of the action of the effector on the receptor and how is this achieved? A physicochemical analysis based on thermodynamic principles provides the framework for answering these questions, but clear insights into the molecular mechanisms requires detailed studies of the structures and interdependencies of the systems involved. These are the provinces of experimental biology.

EXERCISES

1. A new phage ϕX2001 has been discovered which consists of a circular DNA with six sites for attachment of six (identical) coat-protein molecules P. In the host bacterium, the fully-coated phage is assembled stepwise (and reversibly) and the stoichiometric equilibria can be represented as

$$\text{DNA} + P \rightleftharpoons \text{DNA-P}$$

$$\text{DNA-P} + P \rightleftharpoons \text{DNA-P}_2$$

$$\cdots$$

$$\text{DNA-P}_5 + P \rightleftharpoons \text{DNA-P}_6 \quad \text{(COMPLETE PHAGE)}$$

---

* P. Soares de Araujo, M. Y. Rosseneu, J. M. H. Kremer, E. J. J. van Zoelen, and G. H. de Haas, *Biochemistry* **18**, 580–586 (1979).

(a) If the assembly is stopped before all the DNA has been coated, complete phage and all intermediate species are present simultaneously in equilibrium. Find an equation for B_6, *the moles of complete phage* [DNA–P_6] *per mole total DNA present*, in terms of the *stoichiometric equilibrium constants* and the concentration of free coat protein (P).

Answer: $B_6 = (\prod_1^6 K_i)(P)^6/[1 + (\prod_1^6 K_i)(P)^6]$.

(b) Show that if the intrinsic binding affinities for each site are identical and unchanged with increasing occupancy, then,

$$B_6 = \frac{K^6(P)^6}{[1 + K(P)]^6},$$

where K is the intrinsic binding constant.

(c) Suppose the site binding constants $k_1, \ldots, k_6$ were each different (e.g., because the base composition at each site is different) but constant. Prove that the first stoichiometric binding constant K_1 is given by

$$K_1 = k_1 + k_2 + \cdots + k_6.$$

2. The apoprotein form of transferrin can bind two moles of Zn^{2+} ion. In a particular buffer, W. R. Harris [*Biochemistry* **22**, 3920 (1983)] has found that binding data lead to values of the stoichiometric equilibrium constants of

$$K_1 = 10^{7.8}, \qquad K_2 = 10^{6.4}.$$

(a) If the two open Zn^{2+}-binding sites on transferrin are initially identical, find the site binding constant k_1 or k_2 for the binding of the first Zn^{2+} ligand. Calculate also $k_{1,2}$ or $k_{2,1}$, assuming they would be the same.

(b) Calculate the free energy of interaction between the protein binding the first mole of ligand and that binding the second mole.

3. Starting with appropriate equations relating stoichiometric equilibrium constants to the concentrations of (P), (A), (PA$_1$), etc. prove that for a tervalent receptor in which the binding sites have *fixed* but different affinities (that do not change with occupancy by ligand) the following relationships hold:

$$K_1 = k_1 + k_2 + k_3,$$
$$K_1 K_2 = k_1 k_2 + k_1 k_3 + k_2 k_3,$$
$$K_1 K_2 K_3 = k_1 k_2 k_3,$$

4. The enzyme nortestosterone-ketosteroid isomerase is constituted of

Exercises

three subunits, arranged sterically as follows:

(a) Draw a chart showing all possible site binding constants for this enzyme with its substrate molecule.

(b) Derive the relations between the stoichiometric binding constants and the site binding constants if each subunit has a different affinity for substrate but that affinity is constant, that is, does not change when other sites are occupied.

Answer: See I. M. Klotz and D. L. Hunston, *Arch. Biochem. Biophys.* **193**, 314–328 (1979).

Index

A

Acetaldehyde, 35
Acidity, 51, 54
Activated species, 97, 98
Activation energy, 100, 101
Activity, 37, 38, 39
Activity coefficient, 38, 39
Adenosine triphosphate, 40, 52
Algebra
 fundamental theorem, 106
 polynomial, 106
 roots, 106
Algebraic transforms, 114
Archimedes, 74
Arginine phosphate, 57, 58
Aspartic acid, 36
Aspartate transcarbamoylase, 35, 104, 108
ATP, 40, 41, 52, 53, 57, 58, 61, 62
Avogadro, 81

B

Beer's law, 26
Benzoic acid, 47
Binding
 cooperative, 116, 117
 heterogeneous, 117
 ideal, 107, 111, 113, 117
 nonideal, 115, 116, 117

Binding constants
 intrinsic, 107–108
 simulator, 118, 124, 125, 126
 site, 109, 110, 124, 125, 127, 128, 132, 133
 stoichiometric, 104, 118, 119, 126, 127, 128, 132, 133
Binding sites
 ghost, 122, 123, 124, 125
 identical, 111, 113
 real, 123
Binomial expansion, 109
Black, 5
Boltzmann, 81
Bond energy, 51, 52
 phosphate, 59, 61, 62
 thioester, 63, 64, 66

C

Caloric, 5
Catalyst, 27, 95, 99, 100
Centrifugal field, 73, 74
Chemical potential, 18, 20, 27, 37, 52, 57, 71, 86
Chymotrypsinogen, 24, 26
Clausius, 12, 16, 17
Coenzyme A, 59
Concentration cell, 69, 71
Concentration profile, 96, 97, 99

135

Conservation of caloric, 5
Cooperative binding, 116, 117, 129
Cooperative interactions, 116, 117, 129
Coupled reactions, 56, 58
Criterion of feasibility, 18
Criterion of spontaneity, 10, 12, 13, 19
Cytidine phosphate, 30, 31, 32
Cytochrome, 28, 29

D

Denaturation, 24, 25, 93, 94
Diffusion, 74
Distribution of molecules, 80, 82, 83

E

Eadie graph, 114
Eadie–Hofstee graph, 114
Electrochemical cell, 7, 68–71
Electron orbits, 77
Electrostatic interaction, 130
Energy, 3, 4, 5, 8, 10, 11, 12, 16, 20
 activation, 100, 101
 bond, 51
 differentiated, 14
 gravitational, 86
 ionization, 78
 phosphate bond, 59–61, 62
 thermal, 81, 82
Energy levels, 77, 78, 79, 80, 81, 83, 85
 electronic, 78
 rotational, 78, 82
 translational, 78, 81
 vibrational, 78, 82
Enthalpy, 20, 30
 of formation, 31, 47
Entropy, 9, 13, 14, 15, 16, 17, 18, 19, 20, 21, 87, 88, 89, 91, 92, 93, 94
Enzyme, 27, 97, 98, 99, 100
 inhibitor, 100
 kinetics, 95
 substrate complex, 97
Equilibrium, 13, 17, 20
 multiple, 104
 sedimentation, 73–75
Equilibrium constant, 23
 intrinsic, 107
 simulator, 118, 121, 122

site, 109, 110, 118, 120, 121, 122
 stoichiometric, 104, 107, 118, 120, 121, 122
Exhaustion, index of, 13, 15
Extensive factors, 12

F

Faraday, 28, 70
First law of thermodynamics, 3–8
Free energy, 18, 23, 57, 70, 81, 100, 128, 129, 130, 131
 centrifugal, 74
 equilibrium, 22
 Gibbs, 19
 of formation, 32, 47, 48, 49
  table, 33
 of transfer, 46
 standard, 21, 43
 unitary, 39, 47
Fundamental theorem of algebra, 106

G

Ghost site, 122, 123, 124, 125
Gibbs, 19
Gibbs free energy, 19
Glucose phosphate, 24, 57, 65
Glutamine synthetase, 61
Gravitational field, 3, 9
Group-transfer potential, 50, 53, 56, 58, 64

H

Hanes graph, 114
Heat, 5, 6, 8, 12
 frictional, 5
 of formation, 31
Helmholtz, 19
Hemerythrin, 48
Hemocyanin, 48
Hemoglobin, 44, 45, 104
 A, 26
 normal, 26
 mutant, 26
High-energy bond, 51, 52, 59
Hildebrand–Benesi graph, 114
Hydrogen bonding, 90

Index

Hydrophobic bond, 45, 46, 47, 49
Hydrophobic interaction, 130, 131

I

Ice, 11, 21, 87, 90
  melting, 87
Identical sites, 111, 113
Immunoglobulin, 104
Index of exhaustion, 13, 15
Intensive factors, 12
Interaction
  cooperative, 129
  electrostatic, 130
  free energy, 129
  hydrophobic, 45, 46, 47, 129, 130, 131
Intrinsic binding constant, 107–108
Ionization energy, 78

J

Joule, 5

K

Kelvin, 12
Kinetics, 95–102, 107
  enzyme, 95–102

L

Langmuir graph, 114
Ligand, 103–133
Linear transform, 113
Lineweaver–Burk graph, 114

M

Methyl transfer, 66, 67
Molecular disorder, 80, 89
Molecular statistics, 81, 83, 85, 87, 88, 90, 95, 129
Molecular weight, 73, 75, 89
Multiple equilibria, 104

N

Nernst, 71
Nonideal binding, 115

O

Osmotic pressure, 71, 72
Oxidation–reduction, 28, 29, 69
  potentials, 28, 29, 51, 52, 55
Oxygen, 44

P

Peptide, 30, 34, 35
Perpetual motion, 27
Phosphate bond energy, 59, 61, 62
Phosphoglucomutase, 24, 65
pK, 31, 54, 55
Polynomial, 106
Potential
  chemical, 18, 20, 37, 52, 57, 71
  coupled, 58
  group transfer, 50, 53, 58
  oxidation–reduction, 28, 29, 51, 52, 55, 69, 70
  redox, 28, 29, 51, 52, 55, 69, 70
  transfer, 50, 61
Protein
  denaturation, 24, 93
  native, 93
  side chains, 49
  unfolding, 93, 94
Pyruvic acid, 35

Q

Quantum levels, 77, 78

R

Racemization, 34
Receptor, 103–133
Redox potential, 28, 29, 51, 52, 55
Reduction potential, 55
Renaturation, 26
Retinal, 36
Reversible transformation, 13
Rhodopsin, 36
Ribonuclease, 30
Rubber, 91, 92

S

Saturation binding, 105
Scatchard graph, 114, 115, 116, 117, 118

Scott graph, 114
Second law of thermodynamics, 9–17
Sedimentation equilibrium, 73–75
Semi-logarithmic graph, 112, 113, 115, 116
Serum albumin, 104, 128, 129, 130
Simulator binding constant, 118, 124
Simulator equilibrium constant, 118, 124, 125, 126
Site binding constant, 109, 110, 124, 125, 127, 128, 132, 133
Site equilibrium constant, 109, 110, 124, 125, 127, 128, 132, 133
Sites
  ghost, 122, 123, 124, 125
  real, 123
Spontaneity, 10
Standard conditions, 24
Standard state, 38, 39, 42, 61, 63
  rational, 39
  unitary, 39
Statistical thermodynamics, 76, 86
Statistics, 108
  molecular, 81, 83, 85, 87, 88, 90, 94, 129
  of dice, 83, 84
Stoichiometric equilibrium constant, 104, 118, 119, 126, 127, 128, 132, 133
Succinic acid, 41, 42, 43, 44

T

Thermodynamics
  first law, 3, 8
  second law, 9, 13
  statistical, 76, 86
Transfer RNA, 35
Transforms, 106, 113, 114
  algebraic, 106
  linear, 113
Transition state, 95, 96, 97, 98, 100, 101

U

Ultracentrifugation, 73
Unitary free energy, 39, 47

W

Water, 10, 20, 21, 92, 93
  ice, 10
Wolff graphs, 114
Work, 3, 6, 8, 12
  capacity, 14, 15, 17
  electrochemical, 7
  gravitational, 3, 6
  of extension, 4, 7

DATE DUE

C			
MAY 1 0 1994	RETURNED MAY 0 5 1994		
JUL 8 - 2006			
Rapid-12	AUG 2 7 2006		

DEMCO 38-297